THOMAS TELFORD

L. T. C. Rolt was born at Chester in 1910. After his education at Cheltenham College he embarked on an engineering career, until he decided to turn to writing. From childhood he was fascinated by the history of engineers and engineering, and his writing reflected this interest. His first book, *Narrow Boat*, published in 1944, describes a journey along the English canals during the twelve years that he lived afloat. His subsequent biographies of famous engineers, like his writings on railways and motor cars, show his concern to give the story of the Industrial Revolution an imaginative and literary shape.

He was a founder member of the Vintage Sports Car Club and was also co-founder and first Honorary Secretary of the Inland Waterways Association. He founded the Talyllyn Railway Preservation Society, of which he was Vice-President for many years, and he was a member of the Science Museum Advisory Council and Vice-President of the Newcomen Society for the study of the history of engineering and technology. He was a Fellow of the Royal Society of Literature, in 1965 was awarded an Honorary M.A. degree by the University of Newcastle and in 1973 received an Honorary M.Sc. from Bath University. Mr Rolt died in 1974.

Among his many publications, he has written biographies of *Isambard Kingdom Brunel* and *George and Robert Stephenson* (both published in Penguins), and two volumes of autobiography, *Landscape with Machines* and *Landscape with Canals*.

L. T. C. ROLT

Thomas Telford

*

PENGUIN BOOKS

Penguin Books Ltd, Harmondsworth, Middlesex, England
Penguin Books, 625 Madison Avenue, New York, New York 10022, U.S.A.
Penguin Books Australia Ltd, Ringwood, Victoria, Australia
Penguin Books Canada Ltd, 2801 John Street, Markham, Ontario, Canada L3R 1B4
Penguin Books (N.Z.) Ltd, 182–190 Wairau Road, Auckland 10, New Zealand

—

First published by Longmans 1958
Published in Pelican Books 1979

Made and printed in Great Britain
by Hazell Watson & Viney Ltd,
Aylesbury, Bucks
Set in Monotype Bembo

Where these capacious basins, by the laws
Of the subjacent element, receive
The Ship, descending or upraised, eight times,
From stage to stage with unfelt agency
Translated, fitliest may the marble here
Record the Architect's immortal name.
TELFORD it was by whose presiding mind
The whole great work was planned and perfected;
TELFORD who o'er the vale of Cambrian Dee
Aloft in air at giddy height upborne
Carried his Navigable road: and hung
High o'er Menai's Strait the bending bridge:
Structures of more ambitious enterprise
Than Minstrels in the age of old Romance
To their own Merlin's magic lore ascribed.
Nor hath he for his native land performed
Less in his proud design; and where his piers
Around her coast from many a Fisher's creek
Unsheltered else, and many an ample Port
Repel the assailing storm: and where his Roads
In beautiful and sinuous line far seen
Wind with the vale and win the long ascent
Now o'er the deep morass sustained and now
Across ravine or glen or estuary
Gaining a passage through the wilds subdued.

(Lines by Robert Southey inscribed on marble beside
the Caledonian Canal, Clachnaharry, Inverness.)

ACKNOWLEDGEMENTS

THE Raeburn portrait of Telford, No. 1, is reproduced by kind permission of the Lady Lever Art Gallery, Port Sunlight, while the cameo portrait, No. 2, is reproduced by courtesy of Mrs. N. S. Telford. Nos. 3, 5, 8 and 9 are reproduced by courtesy of the Council of the Institution of Civil Engineers. Nos. 6, 10, 11, 12, 13 and 14 are from photographs supplied by Eric de Maré, Esq., A.R.I.B.A. Mr. de Maré also supplied the reproductions of Nos. 7 and 16. Number 15 is Crown Copyright and is reproduced by permission of the Director of the Science Museum, South Kensington. No. 17 is reproduced by permission of the Trustees of the National Portrait Gallery. No. 4 is reproduced from a lithograph in the Victoria and Albert Museum.

CONTENTS

FOREWORD   11

1. MORNING IN ESKDALE   17

2. LONDON AND PORTSMOUTH   24

3. THE SHREWSBURY DAYS   32

4. STREAM IN THE SKY – THE ELLESMERE
   CANAL   50

5. ROADS TO THE ISLES   74

6. THE CALEDONIAN CANAL   92

7. THE GOTHA CANAL   107

8. THE ROAD TO HOLYHEAD   123

9. THE COLOSSUS OF ROADS   142

10. THE COMING OF RAILWAYS   164

11. THE LAST CANAL   184

12. EVENING IN ABINGDON STREET   198

A NOTE ON SOURCES AND
ACKNOWLEDGEMENTS   207

BIBLIOGRAPHY   210

INDEX   211

ILLUSTRATIONS

1, 2. Two Telford portraits

3. Elevation of the church of St. Mary Magdalen, Bridgnorth

4. The stream in the sky: Pont Cysyllte Aqueduct

5. Cartland Crags Bridge

6. Tongueland, River Dee

7, 8. The Menai and Conway Suspension Bridges

9, 10. The Menai and Conway Suspension Bridges today

11, 12. Iron mastery: Detail of spandrel, Waterloo Bridge, Bettws-y-Coed, and two of the suspension chains, Menai Bridge

13. Over Bridge, River Severn, Gloucester

14. The opening of the Gotha Canal: The Royal Yacht *Esplendian* entering the sea lock at Mem, 26 September 1832

15. Unachieved masterpiece: Telford's design for a new London Bridge

16. Count Baltzar von Platen

17. Thomas Telford in the last year of his life

MAPS

Ellesmere Canal and Connections 1806   51

Caledonian Canal   93

Gotha Canal   108

Birmingham and Liverpool Junction Canal   186

FOREWORD

THOMAS TELFORD, the 'Colossus of Roads' as the poet Southey called him, bears a name that is honoured among engineers the world over. To the general public, however, he is comparatively little known. It is probably true to say that for every ten people to whom the names of George Stephenson or Isambard Brunel are familiar, only one has heard of Telford. Even then the reaction is often no more than a puzzled frown and a question: 'Telford? . . . yes, I've heard of him, what did he do?'

The reason why Telford is forgotten where Stephenson and Brunel are remembered is this. Whereas the latter built their roads of iron, Telford laid his in water and in stone. His achievement was as great as theirs and of equal historical significance, but his life's work ended just as the railway age began. Railways enabled a new industrial society to take another giant stride forward and in the process Telford and his works were eclipsed; his roads fell silent and his canals slowly sank into neglect and decay.

Telford lived long enough to see steam locomotives at work on the first railways but he died unconvinced of the merits of steam power on rails, and this helps to explain his personal eclipse in the later nineteenth century. To a new generation of engineers who could think of nothing but railways such scepticism seemed fantastically short-sighted and reactionary, the more so because it was widely misrepresented. He was alleged to have rejected the idea of steam locomotion outright, whereas in fact Telford's doubts applied only to the railway locomotive. He could not see the virtue in a machine requiring the provision of a costly special track which must be reserved for its exclusive use and by which its range of movement was restricted. He believed that the future lay with the steam road carriage, free ranging over ordinary roads where, as on his canals, the authority owning the road would not monopolise its traffic. In other words he held fast to the traditional conception of the 'common road'.

Telford's vision of a service of steam carriages plying over his great road from London to Holyhead was never realised – crippling legislation saw to that – and the railway engineers were able to prove to their own satisfaction that he was wrong. Yet the traffic which roars day and night along Telford's road today proves that in the long term view the man who pinned his faith to the road of stone was right after all. But for Telford this vindication has been too long delayed. By the occupants of the hurrying cars the name of the man who drove their roads on such easy gradients through the mountains of North Wales or the Highlands of Scotland has been forgotten. Here is some account of how those roads were made.

Telford's roads died suddenly and have as suddenly come to life again. The death of his canals has been slow and lingering with no comparable revival. Across the Welsh Border to Llangollen there runs the first of his canals, the Ellesmere as it was then called. When it was built it excited as much wonder as did the first railways. Never before in history had man created works of such magnitude as the mighty aqueducts which Telford flung across the valleys of the Ceiriog and the Dee. They stand to this day exactly as he built them. Yet as the traffic they carried ebbed away, first to the rail and later to the road, they were forgotten. Today, although the Holyhead road passes hard by them, few travellers even spare them a glance.

Not very far away, to the south-east on the borders of Staffordshire and Shropshire, runs Telford's last canal, the Birmingham & Liverpool Junction. A direct water route between the industrial Midlands and Merseyside, it was built to counter the threatened railway invasion. Express trains still thunder down the magnificent ways that Stephenson and Brunel laid for them but this canal of Telford's, although he engineered it at an earlier date upon the same grand scale, has become a lonely, forgotten road. Only the float-watching fishermen and the crews of the long narrow boats, which all too rarely disturb its still waters, know its deep tree-shadowed cuttings and the lofty embankments from which the eye can range far across the Shropshire plain to the Wrekin and the border hills.

Yet the past decade has seen a remarkable revival of interest in the canals. Every year more and more people are forsaking the hurly-burly of traffic-congested main roads to spend their holidays and week-ends travelling by boat along these quiet waterways. In this way Telford's achievement as a canal engineer is being rediscovered and re-assessed. But this new generation of watermen know nothing of the heroic engineering struggles of the men who built the roads they travel. They have become so silent, so peaceful, so much a part of the land-scape that the shouts of the sweating navvy gangs, the clink of picks, the thunder of straining hooves and the rumble of wagons on iron ways sound very faint and far off. On Shelmore Great Bank, for example, Telford, in his old age, fought a battle against misfortune that was every whit as bitter and as protracted as that which Stephenson waged against the rising waters in Kilsby Tunnel on the London & Birmingham Railway. Every railway enthusiast knows something of the story of Kilsby, but who has even heard the name of Shelmore? Such bygone dramas have been totally forgotten, buried deep among the pages of old minute books. Here, an attempt has been made to bring them to life.

Another nickname which Telford earned was 'Pontifex Maximus', for he was one of the world's greatest bridge builders. Probably no other man has ever built so many. We find them especially in Scotland, for it was here, in his own country, that his greatest work was done. No man in his century performed a greater service for Scotland. In the year 1801 when Telford toured the Highlands for the first time he found an almost trackless country of dispirited people for whom no help had come since the final defeat of the Clans at Culloden. Yet when, as an old man of seventy-two, he came back to the Highlands for the last time in 1829, he had given the Highlander harbours, built roads to the Isles and to the farthest north, driven a canal through the Great Glen, and spanned Tay and Spey, Dee, Findhorn, Beauly and Conon with his bridges.

Telford built his bridges both in stone and in iron, and of this last he was the first and the greatest master. No other man has ever handled cast iron with such complete assurance and understanding, his exact knowledge of the capabilities of a new and intractable material enabling him to achieve that perfection of proportion which gives strength the deceptive semblance of fragility. His last masterpieces of bridgework were his suspension bridges at Conway and over the Straits of Menai, and it is only in recent years that the weight of road traffic has become too great for them. At Conway a new bridge has now been built beside his; at the Menai the great stone piers and towers still stand to his memory although the roadway platform and his massive wrought-iron suspension chains were replaced in 1940. It is fitting that this work of reconstruction should have been sympathetically carried out under the direction of Sir Alexander Gibb, the great-grandson of one of Telford's most trusted assistants.

As this story will show, Telford embarked upon his career with the intention of becoming an architect and it was as an architect that he thought of himself until he turned to engineering at the age of thirty-six. He was thus an engineer-architect and brought to civil engineering that architectural quality which it had lacked before and which we find so admirable today. The step from one profession to the other was easily made in those less specialised days and there were others who took the same path although they failed to achieve such eminence. It was a step of great historical importance, for such a diversion of talent into new, adventurous and novel fields may help to explain why the rise of the civil engineer marked the end of the golden age of architecture in England. Telford's scathing comment on contemporary architects 'blundering round about a meaning' has a wide significance.

Telford passed on the architectural tradition to the first great railway engineers, but after them it slowly perished in the jungle of commercial expediency. His other contributions to engineering have proved more lasting. They sprang from his gifts of organisation and leadership and the infallible judgement he displayed in picking the right men for the right jobs and training them to serve his purposes. Although he acquired considerable theoretical knowledge, he always laid the greater emphasis on practical experience and despised the pure theorist. Of his own beginnings as a skilled stonemason he once wrote: 'This is the true way of acquiring practical skill, a thorough knowledge of materials to be employed and, altho' last, not least, a perfect knowledge of the habits and disposition of workmen.' It was this 'perfect knowledge' which enabled him, when he undertook the building of his great aqueducts, to assemble under his command an incomparable team of engineers and craftsmen who followed their leader faithfully from one great work to another until they died. This book is a record of their collective achievement, although Telford's was always the presiding mind.

Telford was the founder of a school of engineers, and some of these men whom he trained became his contractors, commanding labour forces whose strength often exceeded a thousand men. For it was not, as is generally supposed, the railway builders but Telford who first established the system of large-scale engineering contracting. He laid down a system and demonstrated a high-principled and smooth-working relationship between engineer and contractor which was to serve as a model to his successors for generations to come. It is not only as first President of the Institution of Civil Engineers that Telford deserves to be called the father of his profession.

Today we can only marvel at the amount of work Telford managed to cram into one lifetime. His spur was not honour or monetary reward but something which has become extremely rare today – an immense gusto and enthusiasm. His work fascinated him and he devoted himself to it utterly. The more difficult the task the more it appealed to him. When he achieved fame his practice became enormous and no attempt has been made in this book to make a catalogue of every work upon which he was consulted; that task has already been most conscientiously performed by Sir Alexander Gibb in 1935. Here the aim has been to write a more detailed account than has been available hitherto of the major works with which he was most closely associated and at the same time to present a portrait of Telford the man. This second aim cannot be wholly achieved, for despite every effort on the part of the biographer he remains an enigmatic character. His friends seem to have found this tall, massively built Scot with the thick brown curls and strongly moulded features a simple, unassuming, kindly man, humorous, open-handed and open-hearted. This was a façade behind which few, if any, of his contemporaries were ever permitted to pass. He had no wife and he left behind him few truly personal papers so that the clues to his nature are all too scanty. But that the man who devoted so much of his leisure to reading and writing poety was a more complex personality than many of his friends supposed cannot be doubted. But he keeps his secret well. Here then, so far as the man himself and the passage of the years permit, is the story of Thomas Telford, F.R.S., one of the greatest of Britain's engineers.

1958.

I

MORNING IN ESKDALE

On a bank above the Megget Water and in the shade of a single ash tree a few stones mark the site of a shepherd's shieling. Invisible from the stream below, they will soon be altogether lost in the twilight of a new conifer plantation. The little rectangle of stones does not break the ranks of the young trees as they march down the slope, for to the men who planted them it possessed no significance. Today these trees are still no more than knee high so that one can stand upon what was once an earthen floor and look north or south along the narrow valley. It is a prospect that can have changed scarcely at all since John Telford, the shepherd of Glendinning, looked out upon it from this place on a morning in August 1757.

The way in which the hills, Westerker Rig and Effgill, sweep down to the Megget Water is gentle but majestic; majestic in ample curve and fold but gentle because most richly green from base to crest and sewn still with sheep as thickly as a night sky with stars. Northwards, in a shelter-belt of trees, is the farmstead of Glendinning. Beyond the farm the valley floor narrows and rises and the thin ribbon of road falters long before it reaches the Megget's source on that high ridge which thrusts southwards from Ettrick Pen to form the march between the counties of Dumfries and Selkirk. Southwards the little valley with its meandering stream can be followed to its junction with the broader and richer parent valley of the Esk; for from this vantage its cup is seen partly filled by other green hills which stand to the south of the Esk, the Shin and the higher Cauldkine Rig beyond it.

Northwards in the Highlands of Scotland savage screes and the cloudswept darkness of crag and corrie are a perpetual reminder of tragedy and violence. Who can visit Glencoe and not remember the massacre of the Macdonalds? But here in the Lowlands the landscape is so gentle and kindly that to recall that this was once the country of the border reivers requires a considerable imaginative effort. In 1757 it was otherwise. Only fifty years had then passed since the Act of Union had finally extinguished lawlessness on the border and the memory of it was still green. Jamie Telfer of the Border Ballad was a bygone kinsman of the shepherd John Telford and the reiver blood still ran strong. Years later his only son recalled his family's wild past and wrote:

> Alternate plunder mark'd the varying years,
> Each evening brought its triumphs or its tears,
> While Power and Rapine grew from sire to son
> And the song sanction'd what the sword had won.

But now those days were over.

It was on this August day in 1757 – 9 August to be exact – that Janet Telford presented the shepherd with this only child in the tiny cottage by the Megget Water. They named him Thomas. John Telford did not live long to enjoy his son; he died before the year was out. His memorial stands to this day in the long grass of the disused graveyard at Westerkirk in the valley of the Esk three miles from Glendinning. It is as simple as the man himself, a slab of local stone inscribed:

'In Memory of John Telford who, after living 33 Years an Unblameable Shepherd, Died at Glendinning, November, 1757.'

To follow with the finger this incised lettering as the chisel moved in forming it is to perform a magic ritual, to conjure up the past. For the hands that drove the chisel were Thomas Telford's. This was his first work, and for all he knew then he might leave upon the earth no other record of his craftsmanship than this one small stone.

Close beside the grave of John Telford stands an extraordinary building in the form of a miniature rotunda. The work of the brothers Dalziel of Edinburgh it is a masterpiece of the macabre. On the frieze between the fluted Doric columns which flank the doorway and the pediment above are carved the skulls of beasts; the empty niches which flank the columns upon either hand contrive to suggest that they were constructed to house some far more horrifying symbols of mortality and that their vacancy is but temporary. Within the rotunda a continuous frieze of human skulls runs round the wall at waist height. The building is still weather-tight and its iron gate hangs open, but that there is no evidence that any benighted traveller has ever taken advantage of its invitation is scarcely surprising. This is the mausoleum of the Johnstones of Westerhall, a family as long rooted in Eskdale as the Telfords but wealthy gentry when the Telfords were field labourers. It was William Johnstone of this family who was to set Thomas Telford on the road to fame.

A more elaborate example of young Telford's workmanship was a monumental slab set in the wall of the old church of Westerkirk which has now gone. This, with its moulding and coat of arms, commemorated James Pasley of Craig, a member of another long-established local family which played a part in shaping Telford's character and career.

The little shepherd's croft at Glendinning was what we should now call a tied cottage, so Janet Telford had to go. Six months after her husband's death she moved with her infant son to a small cottage at the Crooks which is situated in the Megget valley a mile below Glendinning. By local custom the widow was now generally called by her maiden name of Janet Jackson, and here she raised her son to manhood. They occupied only one of the cottage's two rooms, a neighbour named Elliot living in the other. Life was hard for them and would have been harder still had it not been for the generosity of neighbours, in particular Telford's uncle Jackson who paid the small fees which enabled him to attend the parish school at Westerkirk. Here, among his schoolfellows, Telford made friends to whom he remained faithful throughout his busy life, the most intimate being the brothers William and Andrew Little.

Before he went to school and, later, in the long summer days when the school was closed, Telford worked for neighbouring farmers, bird scaring, herding cattle or living for weeks on end with shepherds in their lonely bothies on the hills. We now think it barbarous that a small boy should have been forced to work in this way, yet this experience, besides making Telford self-reliant at an early age, brought him another priceless gift. If school brought him the gift of good friends, the long lonely days and nights which he spent as a shepherd's boy upon the high hills awoke in him an intense love for his native place. In the shaping of life, of character and destiny, this love of place is of incalculable importance. It can flower mysteriously in the heart in response to the smallest things: the singing of the mountain wind through the bents; the distant bleating of sheep or the cry of a curlew; the first glimmer of the evening star over some moorland ridge night-dark against the sunset. It was a love which Telford never lost. When he achieved fame his visits to Eskdale became of necessity brief and infrequent, but the memories of its green hills and swift falling streams as first seen through the child's clear, wondering eye remained with him, bright and untarnished, till the day of his death. Nowadays, if a man decides to follow the profession of engineer he must usually reconcile himself to urban and industrial surroundings. And if he strays from such surroundings into the landscape he will leave it desecrated by the bleak effrontery of steel and reinforced concrete. It is a strange coincidence that Telford, the man of the hills and the lover of natural beauty, should have been called upon to execute his greatest works in the Highlands of Scotland and among the mountains of North Wales. That these works match the grandeur of their setting, however, is no coincidence for they represent the mature achievement of the eye that was first opened in wonder upon the hills of Eskdale.

When he left school, Telford was apprenticed to a stonemason at Lochmaben, but this new master ill-treated him and after a few months 'Laughing Tam', as he was called at Westerkirk, was back at the Crooks, much to his mother's dismay. It was at this juncture that Janet's nephew, Thomas Jackson, came to the rescue. He had become land steward to the Johnstones of Westerhall and was therefore a man of some influence in the neighbourhood. He persuaded Andrew Thomson, a master mason in Langholm, to take the boy as his apprentice.

Langholm, the hub of Eskdale, is only a small town today and then it was no more than a village, yet the practical experience which Telford acquired under Andrew Thomson, first as an apprentice and later as a journeyman, was remarkably extensive and valuable. This was due to the fortunate circumstance that after the young Duke of Buccleuch succeeded to his family's great estates in 1767 he put in hand a great programme of improvement. Miry tracks were paved, fords were bridged and substantial stone-built farmsteads began to take the place of the older mud walls and thatch. Even in Langholm mud walls were the rule rather than the exception, but now new-cut stone rose in their stead, while to the west of the Esk what was called the new town grew in what had been open fields. To connect this new town with the old Andrew Thomson and his journeyman assistant threw a fine stone bridge across the river.

Samuel Smiles, in his *Life of Telford*, repeats an amusing local story about this bridge. Andrew Thomson was away from home when a great storm set the Esk roaring down in tremendous spate upon the newly completed piers. Andrew's wife Tibby, her fears preyed upon by local wiseacres, became convinced that the bridge would be swept away and ran down the street crying: 'We'll be ruined – we'll all be ruined! Oh where's Tammy Telfer – where's Tammy?' When she had found him, Telford ran with her to the bridge and, finding that all was well, tried to reassure her. But the story goes that Tibby ran on to the bridge and set her back against the parapet as though she would hold it up, a spectacle that made Telford almost speechless with laughter. That laughter echoes down the years to us because the joke has become richer in the passage of time. For the bridge still stands fast today having defied the anger of Esk floods for nearly two centuries. Nor have water or weather yet rubbed out the mason's mark which the young journeyman, proud in the mastery of his craft, cut in its stones. In seasons of spring or summer drought when the Esk runs tamely between wide gravel spits it is possible to walk dryshod under the western arch of Langholm bridge. If he who does so looks diligently he will discover Telford's mark, thus: ⬦ graven upon certain blocks in the western abutment.

Telford's pride was justified, for there is no doubt that by the time he left Langholm he had become an accomplished craftsman in stone. This practical mastery was always a source of great satisfaction to him while it undoubtedly contributed to the respect in which he was held by engineers and contractors who worked under him. They knew they were dealing with a man who had learned his job the hard way, from bed rock, and who could never be fooled by evasions or excuses. Telford went on working in Langholm as a journeyman for some years, possibly for another master, and it may have been during this period that he became the firm friend of a fellow mason, Matthew Davidson, who was to play an important part in his life.

Telford not only learnt his craft in Langholm but discovered there the riches of English literature. This was for him as magical a doorway as that which had opened for him when, as a little boy, he had watched sheep upon the hills. Here he found expressed all that he had felt then but had been unable to put into words. He owed this discovery to a Miss Pasley, an elderly maiden aunt of the Pasleys of Craig who lived alone in Langholm. She had heard from Craig about the young mason and his widowed mother at the Crooks; the story won her sympathy and she invited him to come and see her. Telford had shown a great aptitude for reading and writing when he was at school, but there had been little at Westerkirk for appetite to feed upon. But here, in Miss Pasley's little parlour, shelf above shelf of books met his awed and fascinated gaze; here were riches such as he had never dreamed of. And little Miss Pasley, touched by his eagerness and wonder, gave him the freedom of her shelves.

One of the first books he borrowed from her was an edition of *Paradise Lost*. He treasured it like a child with a secret hoard until his work was done and he was able to walk alone on to his beloved hills. Then he sat down to read and was at once caught up and swept away by the surge of that tremendous poem. The effect of the Miltonic line upon an ear naturally tuned to poetry but utterly inexperienced must indeed have been overwhelming. Telford himself could never describe his feelings on this first reading of Milton. He could only say: 'I read and read, and glowred; then read and read again.' So was born the second love which was to endure to his life's end. No matter how busy he might be he never afterwards denied himself the solace of books.

Telford left his native Eskdale for the first time in 1780. Much as he loved it there was no scope for his ambition in Langholm and he was no longer content with the eighteenpence a day which was all he earned as a journeyman. So he walked to the city which, in all Scotland, offered the greatest scope for his craft – Edinburgh. The Nor' Loch (where Waverley Station now stands) having been bridged, drained and filled, the New Town, Princes Street and the noble Georgian streets and squares beyond it, were rapidly covering the green fields above the northern slope of the valley. What work Telford did during the twelve months or so which he spent in Edinburgh we do not know and he was himself silent upon the subject. No doubt a sufficiently diligent search would reveal a specimen of his mason's mark. But whatever work it was it would undoubtedly be upon a grander scale than anything he had previously experienced. So he was a wiser man and could consider himself a fully trained craftsman by the time he returned to the cottage at the Crooks at the end of 1781.

When Samuel Smiles visited Eskdale he asked a local inhabitant what work would be found for the numerous children he saw in the villages. 'Oh, they swarm off', was the reply. The time had now come for Telford to leave the hive for good. The Edinburgh visit had been merely a short trial flight. The real leavetaking came at the end of January 1782 when, like Dick Whittington, Telford resolved to go to London to seek his fortune. A journey to London from the Lowlands of Scotland in the eighteenth century was no small undertaking for a young man of very little means. But once again good fortune and good neighbours seemed to conspire to help him. It so happened that Sir James Johnstone of Westerhall wanted to send a horse to a member of his family in London. As soon as he mentioned this to his steward, Thomas Jackson, the latter realised that his cousin's transport problem was solved. So away went 'Tammie Telfer', now a tall, powerfully built young man in his twenty-fourth year, trotting steadily down the long road through Carlisle to the south, wearing a pair of buckskin breeches borrowed from his cousin and with all his worldly possessions in the bundle tied to his saddle. It was his first long journey but it was the beginning, had he but known it, of a long lifetime of journeying.

2

LONDON AND PORTSMOUTH

MANY of Telford's old school-fellows and contemporaries 'swarmed off' long before he did, with the consequence that Eskdale no longer had so many ties to hold him. His closest friend, Andrew Little, had gone to Edinburgh to train for the medical profession and had then shipped as a surgeon aboard an East Indiaman. But shortly before Telford left for London, Little had returned to his native valley. He had been struck blind by lightning when his ship was caught in a violent storm off the coast of Africa. Unable any longer to follow his old profession, soon after his return Little took charge of the school at Langholm. He continued to teach there until his death in 1803 and the letters which Telford wrote to him provide the most revealing source of information we have concerning his doings, his thoughts and ambitions during the first important years which set him on the road to fame.

He had no sooner reached London than he wrote home on 12 February to Andrew with messages for his mother and the friends he had left behind: to William Little who would read his letter to his blind brother; to 'that glutton boy John Elliot [who] will have nobody to keep him in countenance now; tell him if he'll come here that I'll fight him at Hyde Park Corner. Tell Matthew Davidson that I enquired at Carlisle for his Glass but the man had none. My compliments to Jennie Smith tell her she is a Canterrin sort of a lassie . . .' If Jenny had lost her heart to Tammie Telford she was doomed to disappointment. From now on Telford was to deny himself the company of women and even the comforts and the solace of a settled home. It has often been loosely said of a man that 'he lives for his work' but in Telford's case this was literally true. As one of his friends would say of him after his death: 'he lived like a soldier, always on active service.' It would be easy but quite wrong to infer from this that the mature Telford must have been a cold, hard, puritanical and unsympathetic character utterly consumed by a ruthless ambition. Ambitious he certainly was with a tremendous confidence in his own powers, but it is clear that his friends found him a likeable character, a modest, quiet-spoken man with a humorous twinkle in his eye and an extraordinarily generous nature. This may seem contradictory but it is not. Conceit, aloofness, arrogance and self-assertion often spring from secret self-doubt and lack of confidence. The confidence of true greatness is instinctive rather than conscious and is never assertive. In Telford's case, too, his nature was sweetened by those three loves which he never denied himself: his friends, his books and the memories of his native hills.

On his arrival in London, Telford soon got his foot upon the first rung of the ladder thanks to the good offices of his old friend Miss Pasley of Langholm. She had armed him with a letter of introduction to her brother John Pasley, an influential London merchant. He in turn introduced him to two of the greatest architects of the day – Robert Adam and Sir William Chambers. As Telford discovered for himself, these two men were completely contrasting and indeed anti-pathetic personalities notwithstanding the fact that they had for many years shared the Royal office of 'Architect of the Works'. Telford described Sir William Chambers as 'haughty and reserved' and Robert Adam as 'affable and communicative'. He added acutely: 'a similar distinction of character pervaded their works, Sir William's being stiff and formal, those of Mr. Adam playful and gay.' This was perfectly true. Whereas Robert Adam had brought to architecture a new and light-hearted elegance and grace by introducing and adapting to new purposes ornaments borrowed from a wide variety of classical sources, Sir William Chambers clung inflexibly to the cold, mathematical per-fection of the Vitruvian orders as interpreted by Palladio, and he dis-missed the innovations of Adam as so many 'affectations'. If the meticulous proportions of the designs of Sir William Chambers be likened to a Bach fugue, then the work of Adam has its voice in the music of Mozart. Of the two, Telford confessed that the personality of his fellow-countryman Robert Adam left the more lasting impression on his mind, but it was Sir William Chambers who employed him.

Telford was set to work on the new Somerset House, squaring and setting the great blocks of rusticated Portland stone. This he did to such good purpose that he was very soon promoted to the rank of first-class mason. He was very proud of this handiwork and whenever, in later life, he had occasion to cross Waterloo Bridge he would point out to his friends the stones which he laid in the south-west corner of the building. But Telford was not content to excel as a working stone-mason. His ambition was soaring far higher than this and he evidently held no very high opinion of the abilities of most of his workmates. In July 1783 he wrote to Andrew Little: "'Tis impossible for me to inform you of much concerning myself at present. I am laying schemes of a pretty extensive kind if they succeed, for you know my disposition is not to be satisfied unless when plac'd in some conspicuous Point of view. My innate vanity is too apt to say when looking on the Common drudges, here as well as other places, "[I am] Born to command ten thousand slaves like you" – This is too much, but at the same time it is too true, for I find the workmen here to be more ignorant than they are in Eskdale Shores – not a Mt. David^n among them ...' To this frank admission of vanity and the desire to achieve prominence an uncannily exact parallel is to be found in the writings of the young Isambard Brunel forty years later. The two men speak in the same voice.

The scheme to which Telford refers was to set himself up as a build-ing contractor in partnership with a fellow mason named Hatton, the only intimate friend he had made among the London workmen. He had formed the highest opinion of Hatton's ability. 'He has been 6 years at Somerset House and is esteem'd the finest workman in London and consequently in England,' he told Little, 'he works equally in Stone and Marble ... cutting Corinthian Capitals and other Orna-ments about this Edifice many of which will stand as a Monument to his honour; he understands drawing thoroughly and the Master he works under looks on him as the principal support of his business but I'll tear away that Pillar if my schemes succeed and let the Old beef head and his puppy of an ignorant Clerk try their dexterity at their leisure. ... He ... has been working all this time as a common Journeyman con-tented with a few shillings pr week more than the rest ... but I believe your uneasy friend had kindled a Spark in his Breast that he never felt before.' 'There is nothing but just having a Name here to make a for-tune ...' declared Telford optimistically, but he very soon discovered that he needed one other essential requirement – working capital. In spite of encouragement and promises of employment from Robert Adam, owing to lack of means the firm of Telford & Hatton never materialised. In the meantime, however, although he did not realise it,

Telford was already sowing the seeds of his future success by helping one of his countrymen.

Sir James Johnstone of Westerhall was at this time contemplating extensive improvements to his house in Eskdale and arranged, perhaps at Thomas Jackson's suggestion, for his younger brother William, who was in London, to consult Telford about this work. Now this William had made an extremely fortunate match. He had married a Miss Pulteney, niece of the last Earl of Bath, and had changed his name to Pulteney. Through this marriage he succeeded to great estates in Somerset, Shropshire and Northamptonshire and became reputedly the wealthiest commoner in England. He subsequently inherited the Johnstone baronetcy and the estate of Westerhall when his brother died without issue in 1797. 'Mr. Pulteney and I have made *100 alterations* in Westerhall,' wrote Telford to Andrew Little, 'I am sometimes with him twice or thrice daily.' He goes on to inquire about the current cost of materials in Eskdale and asks Little to obtain a quotation for roofing from a firm of local builders. From this it is clear that he was working out a detailed estimate for Pulteney.

For Telford this small commission came at a most fortunate moment. Parliament had reduced their annual grant for the building of Somerset House, the work was slowing down in consequence, and Telford was beginning to fear that he might soon find himself out of a job. Although Pulteney was twenty-eight years older than Telford, it is clear that the two men took an immediate liking to each other. Pulteney was evidently impressed by the way the younger man drew up the plans and estimates for Westerhall, for when that work was done he removed the threat of unemployment by commissioning Telford to plan and superintend a programme of alterations to the vicarage of Sudborough in Northamptonshire. Pulteney was patron of the living of Sudborough and the work was put in hand in anticipation of the forthcoming marriage of his newly appointed vicar, Archibald Alison. As a result of this work the Alison family were added to the list of Telford's lifelong friends and their home, subsequently in Shropshire and finally in Edinburgh, became one of his regular ports of call when he was on his interminable travels.

Work on Sudborough vicarage occupied the winter of 1783–4, and was immediately followed by another and more important commission; this was to superintend the erection of a new Commissioner's house, a chapel and other buildings at Portsmouth Dockyard. The house was to be built to the designs of Samuel Wyatt and was of considerable size and pretension since it was intended for the reception of Royal visitors to the dockyard. We do not know how Telford obtained this appointment. William Pulteney may have exerted his influence on his behalf, but it seems much more likely that Samuel Wyatt himself recommended Telford. Wyatt held the carpentry contract for Somerset House, so he had doubtless become acquainted with Telford and admired his ability. Samuel Wyatt's work has been somewhat overshadowed by that of his more famous brother James, but he was none the less an architect of distinction. Among his major works were Penrhyn Castle for Lord Penrhyn, Hurstmonceux Place, Sussex, and Doddington, Cheshire. More relevant in this context, however, was his close association with Matthew Boulton and James Watt. He had at this time just begun building the famous Albion Mills[1] for Boulton & Watt, while he designed both Boulton's Soho House and Watt's house, Heathfield, Birmingham. It may well be, therefore, that Telford first made the acquaintance of James Watt while he was working for Wyatt. If so it marked the beginning of a long association with his famous fellow-countryman.

Telford's letter to Andrew Little from Portsmouth Dock dated 23 July 1784 begins: 'I have wrote to you again and again and a long time ago, but have never had an answer.' Either these earlier letters were lost or they have been destroyed for this letter is the first news we have from him since the previous summer. He reports that the work is going on briskly and that 'all my groundings are approved of by the Commissioners and officers here, indeed so far, that they would sooner go by my advice than my Master's which is a dangerous point – to keep their good graces and his both, however, I will manage it –' Here he is evidently referring, not to Wyatt, but to the Admiralty Surveyor, J. Marquand, whom he had hopes of succeeding for he adds later: 'I have a chance by and by to be employed as principal Surveyor, for all the Officers here would greatly prefer it – but these things must come in course.'

The next letter from Telford to Little is dated 1 February 1786 and gives us a very complete picture of Telford's life at this time. It also

1. The first large steam flour mill in the country. John Rennie was responsible for the milling machinery. It was destroyed by fire.

reveals how far he had already risen above the rank of working stone-mason and how zealously he was educating himself. 'You ask me what I do all winter – I rise in the morning at 7 °Clock and will continue to get up earlier till it come to 5. I then set seriously to work to make out Acct^s, write on business, or draw till breakfast which is at 9 – next go to breakfast and get into the Yard about Ten. Then all the Officers are in their Offices and if they have anything to say to me or me to them, we do it. This and going round amongst the several Works brings My dinner time, which is about 2 °Clock; an hour and a half serves this and at half after 3 I again make my appearance when there's generally something wanted and I again go round and see what is going on – and draw till 5 then go to Tea till Six – then I come back to my room and write, draw or read till half after 9 – then comes Supper and Bed Time. . . . My business requires a great deal of drawing and writing – and this I always take care to keep under by having everything ready. Then, as know-ledge is my most ardent pursuit, a thousand things occur that would pass unnoticed by good easy people who are contented with trudging on in the beaten Path; but I am not contented unless I can reason on every particular. I am now very deep in Chemistry. . . . I have a Noble Room in the New House which is sacred as the Sanctorum of a heathen Temple – I have plenty of fire and candle allowed me and often my dear Andrew do I wish to have you along with me. . . . You (would) find And^w that I am not Idle but much in the old Way of stirring the Lake about one to prevent it from stagnating. . . . I am powdered every day, and have a clean Shirt 3 times a Week; Portsmouth is what you have said of it, but what's that to me. The Man that will neglect his Business, and suffer himself to be led by the Nose deserves to be P-x'd. ! . . . I wish And^w that you saw me at the present instant surrounded by Books, Drawings, Compasses, Pencils and Pens; great is the confusion but it pleases my taste and *that's enough.*'

Telford's next letter to Andrew from Portsmouth in 1786 shows that he was not too preoccupied with work and study to forget his friends in Eskdale or to forego his literary pleasure. Poetry was his par-ticular delight. He combed the reviews for notices of new and promis-ing work and was himself writing poetry though it is clear that he found it much harder to shape a stanza than a block of granite. He reflects sadly upon the fact that for him, pledged as he was to follow his active career wherever it might lead, life must of necessity be a perpetual farewell. 'We are so tossed to and fro on this bustling Stage,' he writes, 'that no sooner have we got acquainted with a valuable friend than inevitable necessity divides us from that enjoyment and sends us from him ... to play a part on some opposite Quarter of the Stage where a fresh set of Actors present themselves.' He goes on to recall with deep affection the friends of his boyhood, then to acknowledge how inevitable it was that they should be exiled from their native place and from each other: 'We must be separated; we are all born in that situation which enforces an activity to procure a decent subsistence and born where there was no field to act in, therefore torn asunder, divided and scattered over the face of the Earth; we range in various regions for a scanty supply while chance gives others ease, affluence and abundance ...' But he concludes by quoting proudly:

> But yet I would not have a Slave to till my ground
> To carry me, to toss me while I sleep
> And tremble when I wake, for all the wealth
> That sinews bought and sold have ever earned.

'These lines', he explains, 'are taken from a Poem published, or at least reviewed, last month. I have not seen the reviewers so warm in the praise of any late production: it is called *The Task* and the Author's name is Willm Cowper of the inner Temple. ...' Telford continues by commenting upon the poetic efforts of some of his Eskdale friends and then adds the following: 'Eskdale seems to be the soil for poetasters – not that I mean to include young Master Scott in that description; his performances are wonderful; why, Pope was only a fool to him! The luxuriance of his genius will produce Laurels and Bays so rapidly that the field will be overspread ... That immense genius can grasp Systems by intuition ... Altho' my mind houses more than Poems ... yet I will thank you for telling me from time to time what you know of him. If you can procure a Specimen we will be better able to judge of the Phenomenon.'

Who was this prodigy of whom Telford thought so highly? This is an excellent example of the kind of question, fascinating but not strictly relevant, which can sometimes arise, beckoning like a siren, to tempt the student of history away from his plotted course. If 'Master Scott' was indeed Walter Scott, who was fifteen at this time, then Telford's words are prophetic and he must be credited with considerable literary discernment. Walter Scott was descended from the border family of Scott of Harden, and there is no reason why he could not have visited Eskdale in 1786. As against this, however, neither Scott himself nor his great biographer, Lockhart, makes any positive reference to such a visit, while it was not until 1799 that Scott's first original work appeared in print. But whoever 'Master Scott' may have been, how exciting it would be to discover an example of the work which so impressed Telford and his contemporaries.

Telford concludes this long letter to Little by saying that his work at Portsmouth should be completed by the late autumn and that although he had no clear idea as to what he would do next he expected to return to London. If Telford did return at all to London from Portsmouth his stay was very brief because at this juncture William Pulteney stepped into his life again. Pulteney had become Member of Parliament for Shrewsbury. The Castle of Shrewsbury, derelict for many years, formed a part of his Shropshire estate and as a result of his election he resolved to occupy it. Telford found himself ordered down to Shrewsbury to superintend a thorough renovation of the building and he was installed at the Castle before the year 1786 was out. It would seem that the Portsmouth job had proved, after all, to be rather a blind alley, so Telford must have wondered what the future had in store for him now as he rode westwards towards Wales through the short days of the dying year. It was a journey into fortune had he known it, for it was upon the marches of Wales, amid the hills of that other border country, that he would first win national fame.

3

THE SHREWSBURY DAYS

TELFORD's first letter to Andrew Little from Shrewsbury, written from the Castle at the end of January 1787, reveals him in very high spirits. 'Mr. Pulteney has been gone a fortnight,' he writes, 'and I am again commanding officer in this renowned fortress. . . . I lay the Town and Country under Contribution in every direction (only in the building line) and as this is a kind of Metropolis to Wales it naturally falls under my jurisdiction.' He had good reason to feel elated for he had secured the appointment of Surveyor of Public Works for the County of Salop.

There can be no doubt at all that it was through William Pulteney's influence that Telford secured this appointment. Indeed so obvious was the favour shown by Pulteney towards him that on his own admission Telford became known in the town of Shrewsbury as 'Young Pulteney'. In his climb to fame the patronage of William Pulteney was an immense advantage to Telford, but it would be quite wrong to suppose that without it he might have remained a mediocrity. It is true that in the eighteenth century the power of wealth and privilege was widely abused by the creation of sinecures, but the post of County Surveyor was no sinecure. In the case of Telford, as in the lives of most famous men, it is easy for little men, jealous of their greatness, to maintain that success was due to preferment by lobbying the wealthy and influential and that, given the same advantages, many another could have done likewise. This is part of the fallacious idea that the world is full of 'mute, inglorious Miltons'. The facts are that if ability is great enough it is seldom that a patron will not appear to encourage it and that no patron can do more than create opportunities for genius. Thus in Shrewsbury William Pulteney gave Telford his first great opportunity; only Telford himself could turn it to good account and that he did so simply showed that Pulteney's estimate of his ability was correct.

Although Telford had a deep and entirely sincere admiration for him, William Pulteney was a strange character and cannot have been an easy master to serve. He was an astute man of business who, in public life and in the management of his great estates, used his income of £50,000 a year liberally and wisely. Yet in private life his parsimony was extraordinary; he is said to have begrudged a personal expenditure of £2,000 a year and to have lived so simply that for many years bread and milk was his only food. He could, had he wished, have practised political jobbery on a vast scale because he had at his disposal more pocket boroughs than any other man in England. Yet Pulteney never abused his power, and his reputation in Parliament for honesty and integrity stood very high. Sir John Sinclair said of him that: 'he never gave a vote in Parliament without a thorough conviction that it was right; and men of that description deservedly acquire great influence in a popular assembly. [He was] a quiet man of plain, unadorned language and strong personality whose opinion was always received in the House with respectful attention.' This complete lack of ostentation was the quality which Telford most admired in Pulteney and which he himself emulated. 'His manner is remarkably engaging,' Telford wrote, 'and how comes it to be so? Why, by that plain simplicity and natural ease which ought to be the study of all men ; the moment that is departed from, there is something takes place which is disgusting. Mr. Pulteney is a man who does not court popularity; he is distant, cautious and reserved to mankind in general, but I believe to the few in whom he can confide there is no man more open . . .'

Although Telford once said in jest that 'we fight like tinkers', there is only one recorded instance of a serious disagreement between Telford and his patron and this had nothing to do with business. Politically, Telford was naïve as the few comments he makes in his letters reveal. Before he came to Shrewsbury he admitted that his work and his studies left him no time to think about politics and he seems to have been reasonably content with the *status quo*. But the French Revolution was an event that nobody could ignore and Telford, like many another young man of his day, was carried away by the ideals of the Republic. Pulteney, an older and wiser man, subscribed to Burke's famous defence of the British constitution and the principle of inherited right as opposed to the notion of natural right. Matters came to a head soon after Telford acquired a copy of the first part of Paine's *Rights of Man* upon its publication in 1791. Its instantaneous effect upon his thought is revealed in a letter to Andrew Little. 'I am convinced', he wrote, 'that the situation of great Britain, tho' perhaps not quite so alarming as he [Paine] represents is yet such that nothing short of some signal revolution can prevent her from sinking into Bankruptcy, Slavery and Insignificancy. ... It would require the united abilities, integrity and independence of a hundred *Pulteneys* to purge the bloated mass and restore it to health and Vigour and that, alas! can never be expected. It must make every sincere lover of his Country and the welfare of its inhabitants grieve to reflect with what rapid progress we are hastning towards inevitable ruin. ...' Not content with this, Telford sent a copy of Paine's pamphlet to Langholm under William Pulteney's frank. Its inflammatory effect was such that a group of young men styling themselves the Langholm Patriots took to drinking revolutionary toasts at the Cross in the little town and generally created such a commotion that they were arrested for disturbing the peace and dispatched to the county gaol for six weeks to cool their ardour. When, as a result of this disturbance, Pulteney discovered that his protégé had been using his frank to distribute what he regarded as seditious literature he took the unhappy Telford to task in no uncertain manner. Pulteney forgave him, however, and Telford seems to have taken his advice to leave politics to others and stick to his business. As was the case with Coleridge, Southey and Wordsworth, age and experience combined with the bloody course of events in France to convince Telford that his revolutionary sympathies had been ill-judged and he soon returned to the fold of orthodoxy.

It is important to remember that at the time Telford went to Shrewsbury engineering was hardly recognised as a profession. Great engineering works had already been successfully carried out in England, notably the canals built by James Brindley, but their executants scarcely thought of themselves as engineers. Brindley was a millwright by trade and the word 'engineer' was little used except in the military sense. John Smeaton was the first Englishman to adopt the French mode and describe himself as a Civil Engineer. He it was who founded the Society of Civil Engineers at the King's Head tavern in Holborn on 15 March 1771. But the aim of this Society was social rather than functional and it did not aim to widen the recognition of engineering as a profession. Consequently architecture continued to be the profession which, above all others, attracted ambitious men who longed to excel in the field of applied art. Inspired by the example of Chambers, Adam and Wyatt, Telford's ambition was to make a name for himself as an architect and a great deal of his self-education while he was at Portsmouth and throughout his Shrewsbury period was directed to this end. The extracts which he laboriously copied into his Architectural Commonplace Book show how conscientiously he studied contemporary authorities on both classical and Gothic architecture: Campbell's *Vitruvius Brittanicus*, Stuart and Revet's *Antiquities of Athens*, Bentham's *State of Architecture*, Tytler's *Progress of Architecture* and Montfaucon's *Antiquities*. The influence of the new romanticism is also apparent in extracts from Thomas Gilpin's *Observations Relative to Picturesque Beauty*. Chemistry and other subjects were studied in so far as they could be applied to the art of building.

In Shrewsbury Telford soon found plenty of opportunity to apply the lessons he had learned. After only a few weeks in the town he could write: 'I have drawn 5 or 6 Plans for different alterations at and above the Castle. I have made a Survey and Plan of part of the High Street of Salop, and last Session a vote was made to sweep away the half of one side, on purpose to render it a good and open Street . . . I have drawn a Plan for a private Gentleman to form a kind of square in another Street, and another Plan for a Country house for another Gentleman which will cost about a *thousand*. . . .'

Another work which Telford had to supervise very soon after his appointment as Surveyor was the construction of a new infirmary and county gaol in Shrewsbury. This is of interest because it brought him into contact with John Howard, the great penal reformer. Short and thin with a sallow complexion due, no doubt, to his frail health, Howard had a singularly unprepossessing appearance. He was also a very shy and retiring man who hated any kind of ostentation. Yet this was the man who had successfully attacked the appalling conditions prevailing in prisons, not only in this country but all over Europe. In England he had put an end to the fearful system whereby gaolers were paid by the prisoners in their charge, a system which encouraged horrible abuses and which caused many innocent people to languish in prisons merely owing to their inability to pay off their gaolers' fees. When he came to Shrewsbury in 1788 he was making his last tour of English prisons.

Howard's great spirit quite transcended his physical imperfections and he made a deep impression on Telford. 'You will easily conceive how I enjoyed the conversation of this truly good man,' he told Little, '. . . I consider him as the guardian Angel of the miserable & distressed, travelling over the world merely for the sake of doing good, shunning the society of men and afraid of being taken notice of. . . . He assures me that he was born to be a domestic man and that he hates travelling; that he never sees his country house but he says within himself: "O might I but rest in this spot and never more go three miles from home I should be happy." But he is now entangled in so extensive a plan that he says he is doubtful that he shall ever be able to accomplish it – he goes abroad soon. He never dines, he says he is old and has no time to lose . . .' Howard went abroad in July of the following year and never returned. While nursing a victim of camp fever he caught the infection himself and died at Cherson on the shore of the Black Sea in January 1790.

Plans for the county gaol had originally been prepared by J. H. Haycock, but they were now much altered and improved by Telford as a result of his discussions with Howard. Another prison reform which Howard initiated was the idea that prisoners should be set to do some form of useful work and this proposal was also adopted at Shrewsbury. Gangs of prisoners were sent out to work as labourers on a number of Telford's undertakings in the neighbourhood, including the excavations which he carried out at Wroxeter.

Telford's excavation of the Roman city of Uriconium was a pioneer work of its kind. Archaeology as a scientific study did not exist and very little practical exploration had been done. Certainly it would be hard to find many earlier examples of excavations so extensive as those carried out at Uriconium. The methods employed would doubtless horrify a modern archaeologist because the motive was different. Telford was not concerned laboriously to piece together the evidence of the way of life of the Romano-British inhabitants of the lost city. He embarked upon the work as an architect whose curiosity had been fired by his study of classical architecture. In this we probably have the key which explains how archaeological research in this country began. However destructive its results may have been by modern standards, at least it possessed a negative virtue in that it put a stop to the wholesale quarrying of ancient sites for building material. To read Telford's

account of Uriconium is to marvel that any relics of Roman civilisation
should have survived at all in this country. 'The Farmers never harrow
their ground,' he writes, 'but they find ancient Roman coins and other
pieces of antiquity. They know where there are Ruins underneath by
the Corn being scorched in dry weather, so that when they wish to
Dig for Stones they put down a mark on the place until the Corn is off
the ground, when on removing the Soil they are sure of finding something or another. At this time they wanted Stones to repair a black-
smith Shop, and uncovering one of these places in order to procure
them, they discovered a Bath plastered with Roman Cement, a floor
paved with large Tyles and some Pillars.'

It was at this juncture that further quarrying was prevented. Telford
informed Pulteney, who was Lord of the Manor, and he directed
Telford to continue the excavation at his expense, 'in order that men
of learning might satisfy their curiosity'. Telford did so and discovered,
in his own words, 'a dressing room, a Cold Bath, a hot Bath, a large
Sudatorium, four Tessellated floors and places with pillars of Tyles'.
He wrote a monograph on the excavation and prepared ground plans
and drawings of the buildings discovered.

Of much greater importance where Telford's reputation was concerned was the extraordinary affair of St. Chad's church, Shrewsbury,
which occurred while the work at Uriconium was going forward in
the summer of 1788. As a result of complaints that their church roof
was leaking the churchwardens of St. Chad's sent for Telford and
asked if he would examine the church, make a report, and estimate the
cost of repair. Telford soon found that the leaking roof was the least
part of the trouble. Like all medieval buildings the church had been
built without deep foundations and as a result of some injudicious
gravedigging too close to the walls the structure was subsiding, the
north-west pillar of the tower in particular. Large cracks in the walls
revealed the gravity of the situation only too clearly. Telford reported
upon this alarming state of affairs to the next meeting of the Parish
Vestry which was held in the church. He told them bluntly that it was
pointless to think of repairing the roof until emergency measures had
been taken to secure the walls. This was greeted with a mixture of
indignation and ridicule by the worthy churchmen who assured him
that the cracks which so alarmed him had been there since time imme-
morial. They also made some pointed remarks about the tendency of

professional men to make jobs for themselves. At this Telford lost his patience and left the meeting. His parting shot was to suggest that if they proposed to continue their deliberations much longer they would be wise to adjourn to the churchyard before the church fell down on their heads.

Having dismissed Telford the vestry engaged a local stonemason to underpin the north-west angle of the tower. Nemesis followed more swiftly than even Telford himself could have expected. Three days later, in the early morning, the stonemason was waiting in the church porch for the sexton to come with the key when the church clock began to strike the hour. The effect of the Israelite trumpets upon the walls of Jericho was not more startling. At the very first stroke of the bell the entire tower collapsed with a tremendous roar and crashed through the roof of the nave, demolishing completely the whole of the northern arcade. It formed, remarked Telford dryly, 'a very remarkable magnificent ruin'. This demonstration of professional prescience considerably enhanced 'Young Pulteney's' reputation in the eyes of the inhabitants of Shrewsbury.

St. Chad's church was rebuilt to the designs of George Steuart[1] and there is no evidence to suggest that Telford was consulted about the work. His only known church work in Shrewsbury is the pulpit at St. Mary's. His first opportunity to try his hand at ecclesiastical architecture occurred elsewhere in the county – at Bridgnorth where he designed the new church of St. Mary Magdalene in High Town. No one could claim that this church is a great work, but today few people could be found to agree with Samuel Smiles who dismissed it in these words: 'Telford's design is by no means striking . . . A graceful Gothic church would have been more appropriate to the situation and a much finer object in the landscape; but Gothic was not then in fashion – only a mongrel mixture of many styles, without regard to either purity or gracefulness.' Thus Smiles, writing from that eighteen-sixties nadir of

1. Steuart's St. Chad's with its circular nave 100 ft. in diameter provoked strong feelings. 'This preposterous structure' wrote Robert Southey in 1839, 'with its large round body and its small head has been compared to an overgrown spider. Mr. Telford may have beheld the new church of St. Chad's with some advantage, inasmuch as he saw in it everything that ought to be avoided in church architecture.' On the other hand a modern writer (Mr. M. Whiffen, *Stuart and Georgian Churches*, 1948) has called it 'one of the most boldly conceived buildings of the whole Georgian epoch'. John Simpson (see p. 47) was the builder.

taste against which Morris rebelled, dismissed the age of Adam, Wyatt and Chambers. It is a strange comment on the vagaries of fashion and a salutary reminder that there are no absolute standards of critical judgement in such matters. Telford's next biographer, Sir Alexander Gibb, evidently did not consider St. Mary's, Bridgnorth, worth mentioning, while Telford himself, in his autobiography, was very modest about it. He described it as 'a regularly Ionic interior, a Tuscan elevation and a Doric tower'. 'Its only merit', he added, 'is simplicity and uniformity.' These words were written at the end of his life. In a letter, now, unfortunately, badly mutilated, written to an unknown correspondent at the time the church was building, he describes the ideas behind his design in detail and with more enthusiasm. It is obvious from this that he approached the problems of ecclesiastical architecture in a severely practical way. He thought that the orthodox form of nave and chancel, long, narrow and lofty, tended to structural weakness and was also acoustically wrong. He also disliked the conventional cleres- tory because, he said, 'two rows of windows convey the notion of there being two heights of apartments and . . . the two rows . . . and the other small divisions become an offensive number of trifling parts'. 'In order to avoid these faults in this Plan & Elevation the body of the Church is brought forward distinctly, the plain Order reaches the whole height of the side Walls, and instead of ten small Windows there are three very large ones on each side. By this means it is hoped that the attention will be drawn to the body of the Church which is here meant to appear as one great and undivided apartment.' In order to secure maximum audibility for the greatest number the body of the church was almost square in plan and the chancel (later rebuilt by Blomfield) was short and wide. According to Telford there was no east window in his original chancel, an altar piece occupying the whole of the end wall. This is confirmed by the plan in the Telford *Atlas*.

In the design of the tower Telford's aim was again that of simplicity 'instead of several sham ranges of different and expensive Orders of columns'. The effect of all this is certainly austere, but there is a splendid spaciousness about the interior of St. Mary's while in the opinion of this writer, Telford's tower makes a most effective and entirely appro- priate end-stop to the vista down the gracious, tree-lined street of Georgian houses which leads to it. Because this street pre-dates the church, the credit here is due to Telford.

On the subject of the strange little octagonal church at Madeley[1] which was built to Telford's design and under his direction between 1792 and 1795 we have no word from him except that it was 'peculiar', but it is obvious that it was another application of the same idea of a church as 'one great and undivided apartment' and of the same practical consideration of improving audibility.

It was while Bridgnorth church was building and Madeley, if not begun, was certainly on the drawing-board, that Telford set out upon an architectural and artistic pilgrimage. His subsequent letter to Andrew Little[2] in which he described this experience is of considerable interest. His first objective was Bath and it seems clear that he went there to inspect building works which Pulteney had in hand. For a year previously he had written in a letter to Little: 'Do you know of any good Wrights about Eskdale who would be willing to go to Bath where they would have good encouragement? I could get 3 or 4 employed, but should wish them to be steady and good hands.' Matthew Davidson was among the Eskdale men who responded to this invitation. As to the nature of this work at Bath we are given no clue but it seems possible that Telford had a hand in it. This is certainly hinted in his trenchant comments on the city. He wrote: 'I think I have told you that Modern Bath has been created by a Mr. Wood, an Architect, a man of very superior talents to whom, if I will, I hope to do justice. Since his time, altho' the rage for Building has been unbounded, yet there has none inherited even a portion of his Genius. I will not even except their present Surveyor who is sinking fast into oblivion. He has lost, or rather not succeeded, in the finest attempt which the World ever afforded of finding his fame above that of any other man as an Architect.

'In Lady Bath's new Town every circumstance was most fortunately combined – the whole the property of one person, the greatest plenty of beautiful Material at the cheapest rate in the World, a great demand for every species of Buildings supported by a great, bold and enlightened employer capable of comprehending the finest and most extensive schemes and the whole growing in the Bosom of Wealth. I know of no instance in Ancient or Modern History of the conjunction of so many favourable circumstances and yet I am sorry to say that I shall be able only to allow that there have been Streets and Squares

1. The present chancel was added in 1910.   2. Dated March 10th 1793.

and Crescents and houses which a demand has forced, but alas – the Architect is not to be found. In the late improvements of the old Town the same hand is blundering round about a meaning tho' he might have far excell'd the Bath of Dioclesian or any of the Roman Works. . . .'

How many architects since Telford have lamented lost opportunities in just the same way! So far have architectural standards declined since Telford's day that we find entirely satisfying work which he condemned and consider his criticism unjustified. Yet in the rage for speculative building in Bath, then a relatively new phenomenon, Telford appears to have diagnosed acutely the beginning of that decline in standards of design and workmanship which was destined to bring the Golden Age of English Architecture to an end within his own lifetime. One of the contributory reasons for that decline is to be found in Telford's own life story. Incidentally, the surveyor whose work he so roundly condemns must have been John Palmer who succeeded Thomas Baldwin as City Surveyor when the latter was dismissed in September 1792. Both Baldwin and his associate John Everleigh were speculative builders who were bankrupted by the failure of the Bath City Bank at the time of Telford's visit. Their successor, Palmer, was responsible for Lansdown Crescent and St. James's Square. He also took the credit for the Theatre Royal although in fact this design was by George Dance the younger.

From Bath, Telford journeyed to London: 'it was a fine day and evening and we escaped safe, tho' the Collectors had been doing duty on Hounslow Heath', he wrote. Here he lodged with a fellow Scot while he devoted several days to intensive architectural study – to looking at public buildings and poring over books in the British Museum and in the Library of the Antiquarian Society. Then on to Oxford where he explored the colleges and was lost in admiration of the collection of paintings in the Library of Christchurch. Of the colleges he wrote: 'there are many noble structures some few being fine pieces of Gothic Architecture, a great deal of mixed tastes and some little good Grecian Architecture. There are some things by Inigo Jones, and 'twas here that Sir Christ. Wren first tryed his Architectural skill,[1] and lately there have been several things done by Mr. Wyatt. But I

1. Telford is here referring to the Sheldonian, but in fact the chapel of Pembroke College, Cambridge, was Wren's earliest work.

think the best pieces of Architecture are those designed by Dr. Aldrich, a Dean of Christchurch.'

This comment of Telford's shows fine discernment and is remarkable when we remember that it was written at a time when architectural taste, as exemplified pre-eminently by James Wyatt, had moved away from the seventeenth-century masters of the English Renaissance. For Doctor Aldrich was a contemporary of Wren and while no comparison can be drawn between a gifted amateur who was responsible only for a few works in his own university and the nation-wide achievements of the great professional, if judgement be confined to the context of Oxford then Aldrich's exquisite chapel at Trinity is incomparable. No doubt Telford also admired the spaciousness, so remarkable for its date, of Aldrich's Peckwater Quadrangle at Christchurch. All Saints church is generally regarded as being Aldrich's least successful work and, unlike Trinity Chapel it has suffered from too much attention from the Victorians. But it is significant that, like Telford's design at Bridgnorth, the interior of All Saints is nearly as broad as it is long.[1]

Of all the pictures which Telford saw both in London and Oxford those he most admired were the Reynolds portraits, the *Medusa* of Peter Paul Reubens and a work by Annibal Carracci at Christchurch. Of the *Medusa* he wrote: 'it actually made me shudder to look at; I wonder how the man durst finish it; he must have had a long handle to his brush.' The Carracci painting depicted the artist's father and brothers in a butcher's shop. 'A fine sett of blackguards they are,' was Telford's comment, 'but tis a wonderful painting.'

From Oxford Telford had to go to Birmingham to visit Egington, a maker of stained glass who had been commissioned to produce windows for his church at Madeley and whom he found at work upon a large window for the Duke of Norfolk showing Solomon entertaining Sheba.[2] His comment on the precocious new industrial town of Birmingham, which was growing so rapidly as it pumped its wares through the arteries of Brindley's canals, was brief and caustic: It is, he wrote, 'famous for *Buttons*, *Buckles* and *Locks* and *Ignorance* and *Bar-*

[1]. H. M. Colvin, in his *Biographical Dictionary of English Architects*, states that only the Peckwater Quadrangle can with absolute certainty be ascribed to Aldrich. It is clear from Telford's comment, that at the time of his visit, these other Oxford buildings were credited to him, rightly or wrongly.

[2]. Egington also produced windows for William Beckford's Fonthill Abbey.

barism. Its prosperity increases upon the corruption of Taste and Morals. Its nick nack hardware is a proof of the first and its Locks and Bars are evidence of the last of the Assertions – and the disposition of the inhabitants confirms the Theory.' Telford had been prejudiced against Birmingham by the recent anti-Republican riot in the town when the mob had sacked the house of Dr. Priestley and destroyed his laboratory. He had sent a long, indignant account of the affair to Andrew Little.

Back at Shrewsbury Telford, as his letters reveal, was handling a steadily increasing volume of business. He was also beginning to prosper and in the autumn of 1789 he was able to send the first of many sums of money back to Eskdale to be divided between his mother and Andrew Little. He did so with the strictest injunction to Little that no one else in Eskdale should know of it. 'I have at present the opportunity of earning money. You have not and I therefore insist on the privilege of sharing a little ... This is not ostentation because I don't wish any body to know of it besides yourself ... To set my mother and you above the fear of want has always been my first object altho' I never have told you so before – and next to that to be that somebody that you have always told me I had the right to be. And I humbly presume that there is a something in it – it may be self confidence but I think I have observed that there has always been a bustle where I was ... You know while I was in Eskdale I was a bone of contention, when I came to London I had nearly caused a confusion at Somerset House; at Portsmo' the Navy Board and Admiralty were engaged and since my arrival here there has been one continual sense of contention ...'

Evidently Telford was still acting true to his maxim of 'stirring the lake to prevent it from stagnating'. He emulated his patron's example by living very simply. 'I have myself for about half a year taken to drinking water only, I avoid all sweets and never eat any nick nacks. I have sowens and Milk for my supper. In short ... I am a queer creature and am not ashamed of being thought singular.' But although he was working so hard and living so abstemiously he never denied himself what he called 'a continual something to keep the Spirit awake'. 'It ever has been and ever shall be my aim to unite those too frequently jarring pursuits, Literature and Business,' he told Little, and in proof of this he frequently sent his friend drafts of poems which he had written for Little's comment and criticism. Among these was a new version of his long poem in praise of Eskdale. He had first written this at Lang-holm, but at Shrewsbury he revised it and had copies printed for dis-tribution among his friends. Other poems of his Shrewsbury period were those in praise of James Thomson and Robert Burns and laments for two dead friends of his youth, George Johnstone and his much loved kinsman and contemporary William Telford. Although he occasionally achieved a felicitous couplet, Telford was no poet. The poetic feeling was there and so was the new romanticism to which Coleridge and Wordsworth gave a voice. But unlike theirs, Telford's gift was far too small and weak to burst the bands of conventional form, stilted diction and threadbare metaphor which make his work, like that of many of his contemporary poetasters, almost unreadable today.

The mediocrity of Telford's poetic performance, however, makes no less significant his avowed intent, as engineer, architect and poet, to combine Literature with Business or, in other words, Art with Science. It was an ambition which he shared with some of the greatest intellects of his age and it was a very laudable one. Adam Smith had realised that the growth of the new industrial society depended upon the principle of division of labour and that this process of specialisation must apply not only to manual tasks but to the pursuit of knowledge. The material advantages which such a highly organised form of society would reap were seen to be illimitable, indeed there is little which modern science has since achieved which these brilliant eighteenth-century minds did not optimistically foresee if only in a glass darkly. They realised, how-ever, that if the perfect society to which they looked forward was to be achieved the specialised lump of mankind must be continually leavened by the philosopher who, in their definition, should be both artist and scientist. His should be the inclusive mind, ranging freely over the whole field of knowledge and using his artistic imagination to knit the innumerable strands together and present the picture whole.

The supreme affirmation of this doctrine was Erasmus Darwin's *The Botanic Garden* with its Fuseli illustrations engraved by William Blake, and it was undoubtedly this poem which fired Telford's dual ambition. He enthused about it to Andrew Little and quoted it *in extenso*, called it 'a very wonderful and masterly performance'. '. . . He has contrived', he wrote, 'to introduce a sketch of almost every thing which is in heaven above, the Earth below, or the Waters under the earth . . . From the formation of Planetary Systems to the blushing beauties of the humble Daisy, nothing escapes his attention.' One can appreciate his enthusiasm. How he must have relished, for example, Darwin's description of the steam engine which begins:

> Nymphs! You erewhile on simmering cauldrons play'd,
> And call'd delighted Savery to your aid;
> Bade round the youth explosive steam aspire
> In gathering clouds, and wing'd the wave with fire;
> Bade with cold streams the quick expansion stop,
> And sunk the immense of vapour to a drop. –
> Press'd by the ponderous air the Piston falls
> Resistless, sliding through its iron walls;
> Quick moves the balanced beam, of giant-birth,
> Wields his large limbs and, nodding, shakes the earth.

The principle of the condensing steam engine has never been more concisely and lucidly expressed, yet, to the modern reader, Darwin's use of classical imagery and metre for the purpose of scrupulously accurate technical description seems bizarre and often bathetic if it is not incomprehensible. Yet, like Telford's poetic efforts, Darwin's great poem is immensely significant. Unless we appreciate the purpose and scope of Darwin's achievement we cannot hope to understand the intellectual climate in which Telford had his being. For although Darwin may not have written great poetry he did achieve the astonishing feat of rendering in imaginative terms which any literate man could then appreciate the sum of human knowledge at that time. It was an achievement which was never repeated or even attempted for, to paraphrase W. B. Yeats: 'things fell apart, the centre could not hold.'

An engineer-architect like Telford and men such as Horace Walpole who moved in the world of art and literature found common ground in their admiration for *The Botanic Garden*. It enlarged their understanding and united them in the inquiring, adventurous intellectual spirit of their age. When we look today at any of the great works which Telford was soon destined to carry out we must remember this unifying spirit in which they were conceived. They were not the products of dead formulae and cold calculation but of a newly won technical mastery fired by artistic imagination and burning enthusiasm; they were chapters in an unique and enthralling adventure story which kindled the imagination of engineer and artist alike. To call such engineering achievements great works of art as Telford's contemporaries did seems to us merely an exaggeration. Yet the tribute was sincere and meant exactly what it said. No arbitrary distinctions could be drawn while Science and Art enjoyed their all too brief love match in the mind.

When Telford was not writing poetry he was reading avidly and when anything impressed him especially he would tell Andrew Little about it. He seems to have been particularly struck by Sheridan's biography of Jonathan Swift; struck not so much by its author as by the towering stature of its subject. 'He was', wrote Telford, '. . . among the Sons of men as the man Gulliver was amongst the Lilliputians. He appears to have been so much superior to the common run of mankind that whenever he chose to exert himself his gigantic powers were irresistible; they ran like a torrent which bore everything before it and which it was even dangerous to attempt to stop.'

Besides his poetry and his reading Telford lost no opportunity of indulging his appetite for other arts. He had no puritanical inhibitions about the theatre and when the players came to Shrewsbury he was their enthusiastic supporter. 'We have the Theatre open for some time in this Town,' he told Little, 'and a Mrs. Jordan from London has played 6 Nights. She is certainly one of the first Comic Actresses of this Age – at least I never saw anything in that way that could come in any degree of competition. Every word, every look is nature and she is thus irresistibly charming. . . .'

Very soon after this he went to a local orchestral concert but concluded sadly that he had no ear for music. 'I might just as well have staid at home,' he wrote. 'It was all very fine, I have no doubt, but I would not give a song of Jock Stewart's for the whole. My coarse organs are not susceptible of those enchanting, sweet, delightful sensations. The melody of Sounds is thrown away upon me. One look, one sentence from Mrs. Jordan has more effect upon me than all the fiddlers in England have. It is certainly a defect – but it is certainly a fact – I have no enjoyment from fine Music. I sit down and am as attentive as any mortal can be, nay endeavour to interest my feelings but all in vain. I feel no emotion unless an inclination to Sleep be reckoned upon. Now Query: Whether this is a natural defect or is it *from ignorance of the subject and want of attention in our Youth.*'

All these different facets of Telford's life and work at Shrewsbury are important for the light they shed upon his developing personality, but from the material point of view of the career which lay ahead of him they could advantage him very little. His most important activity has not yet been mentioned – his bridge building. To Telford at this time bridge designing and building was simply a part of his job as Architect and Surveyor for the county, yet in fact it was his Shropshire bridges which led him away from the architectural career he had planned for himself and into the new profession of civil engineering.

In his very first letter to Little from Shrewsbury Castle Telford mentions proposals and plans for a public bridge which may refer to some rebuilding of the English or Welsh bridges which cross the great loop of the Severn which almost encircles the town. But the first bridge to be built *de novo* to Telford's design was that near the little village of Montford, seven miles from the town, which carries the Holyhead road across the Severn. It was built of local red sandstone quarried at Nesscliffe between the years 1790 and 1792 at a cost of £5,800. This Montford Bridge is noteworthy, not only because it was Telford's first bridge work but because to superintend its construction he sent for his old friend and workmate Matthew Davidson of Langholm and employed under Davidson a mason from Shrewsbury named John Simpson. This marked the beginning of a lifelong business association between Telford and Davidson. Simpson, too, was to be connected henceforth with Telford's works until his death.

In his autobiography, Telford estimates that in five years (1790–96) he was responsible for building no less than forty road-bridges in Shropshire. In addition to these many small works were the Severn bridges at Buildwas, Bridgnorth and Bewdley which were completed to his designs and under his direction between 1795 and 1798. Severn was a river with which Telford had a lifelong acquaintance, for by the end of his life he had been responsible for no less than six out of the thirteen road bridges which span the river between Shrewsbury and the sea, only one of which has since been replaced.

Exceptionally severe storms during the winter of 1795 brought Telford much of this work. Swollen streams along the Welsh March swept away their bridges wholesale and with their aid Severn rose in one of those swift and deadly floods for which she is notorious. Buildwas bridge was carried away; Bridgnorth and Bewdley badly damaged. Telford called it 'a greater inundation than has ever been known in England'. 'Much damage and injury has been done by the various rivers,' he wrote, 'and Severn has not been behindhand ... I have now before me a Plan for one over the Severn at Bewdley in Worcestershire which I have just prepared, and I am likewise drawing one for the town of Bridgnorth, in short I have been at it night and day...'

Of Bridgnorth he tells us no more, but he subsequently celebrated the completion of Bewdley bridge in the autumn of 1798 in these words: 'The dry Season ... has enabled us to raise Bewdley Bridge as by enchantment. We have thus raised a magnificent Bridge over the River Severn in one Season. Which is no contemptible work for John Simpson and your humble Servant. ... John Simpson is a treasure of talents and integrity ... and he has now all the works of any magnitude in this great and rich district. Telford's self-satisfaction is pardonable for Bewdley is indeed a magnificent bridge and worthy of the town's splendid Georgian waterfront, as anyone may see to this day. Built of stone from Arley Quarry four miles upstream and costing £9,000, it is perhaps the finest example of Telford's early work in stone. Of greater significance for the future, however, was Telford's new bridge at Buildwas, a bridge since replaced.

A little distance below Buildwas, at Coalbrookdale, there stood, and indeed still stands, the famous iron bridge, the first in the world. A gifted Shrewsbury architect, Thomas Farnolls Pritchard, produced designs for it but died in October 1777 the month before building began. The bridge sections were cast by the great Abraham Darby III in his Coalbrookdale Foundry nearby, where in 1709 his grandfather, Abraham I, had pioneered the smelting of iron with coke. Thomas Gregory, the foreman pattern maker, also played a key role in the project. It is easy to understand how this example of the use of an entirely new material in bridge building would fascinate Telford when he moved to Shropshire in 1787 and how he would seize upon the destruction of the old Buildwas bridge in the sensational floods of 1796 as a heaven-sent opportunity to experiment himself with iron and see if he could not improve upon the prototype downstream. Not that a desire to experiment was Telford's only motive in designing the bridge in iron. By using this material he could evolve a single-span bridge less obstructive to the river navigation and less liable to be damaged by flood. He designed a much lighter and flatter arch than that at Coalbrookdale in order to resist the tendency of the abutments to slide inwards and so distort the arch as had happened in the case of the latter. The graceful Buildwas span of 130 feet exceeded that of Coalbrookdale by thirty feet yet it was half the weight. Its cost was only £6,000. Buildwas bridge was a worthy 'prentice work of the man who was soon to become the greatest of all masters in the structural use of cast iron.

By the time these Severn bridges were building Telford was engaged in matters of far greater moment. In September 1793 he had become 'general agent, engineer and architect' to the Ellesmere Canal Company which had been formed to connect the rivers Mersey and Dee with the Severn at Shrewsbury. This was an appointment as important to Telford as was the younger Brunel's appointment as engineer to the Great Western Railway Company forty years later. It changed the course of Telford's life and brought him a national reputation, not as an architect but as an engineer.

4

STREAM IN THE SKY – THE
ELLESMERE CANAL

IT was the dream of James Brindley, the father of canals in Britain as George Stephenson was the father of railways, to unite by lines of inland navigation the four great rivers: Mersey, Trent, Severn and Thames. The whole of this ambitious scheme was surveyed by Brindley and his assistants but he did not live to see it completed. The first of these inland waterway routes to be promoted were the Grand Trunk and the Staffordshire & Worcestershire Canals. The Acts authorising their construction both received the Royal Assent on 14 May 1766. The Grand Trunk was projected to run from the Bridgewater Canal at Preston Brook to the river Trent at Wilden Ferry near Derby and soon became known as the Trent & Mersey Canal. The Staffordshire & Worcestershire Canal was to branch from the Trent & Mersey at Great Haywood in Staffordshire and extend southwards to join the Severn at Stourport. By means of these two waterways, therefore, three of the four rivers would be united, and when Brindley died in September 1772 both had been completed with the exception of his great tunnel which was to carry the Trent & Mersey under Harecastle Hill near Kidsgrove to the north of the Potteries. This was not finished until 1777 after eleven years' labour.

The long canal line to the Thames at Oxford took much longer to complete. The Coventry and Oxford Canal Companies whose main lines together formed this through route southwards from Fradley Junction on the Trent & Mersey, both got into financial difficulties long before their works were completed. The northern part of the Coventry's line was finally completed jointly by the Trent & Mersey and the Birmingham & Fazeley Companies, while the Oxford Canal was forced to halt at Banbury for eight years until sufficient capital was forthcoming to enable it to resume its southward march down the Cherwell valley to the Thames. As a result, it was not until January

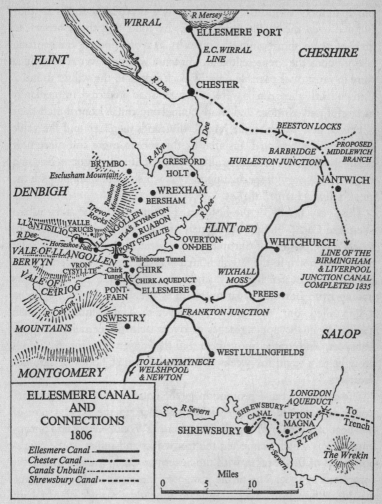

ELLESMERE CANAL
AND
CONNECTIONS
1806

Ellesmere Canal ————
Chester Canal —·—·—·—
Canals Unbuilt ·············
Shrewsbury Canal ·——·——·

Miles

0   5   10   15

1790, nearly eighteen years after Brindley's death, that the first boat loaded with Midlands coal sailed on to the Thames at Oxford to complete his great scheme and open up the first inland water route between the industrial Midlands and London.

On the map of England this early canal system of Brindley's appears roughly in the shape of a St. Andrew's cross and this 'cross' as it is often called, was the skeleton about which the whole network of inland waterways shaped itself. Brindley's canals were narrow, shallow and extremely tortuous; wherever it was practicable to do so they avoided even the smallest earthworks by following the contours. In spite of these shortcomings, however, as a method of transporting heavy goods they represented an immense advance over roads which were in most cases passable only by pack-horses in the winter months. Consequently these canals were, as Brindley had predicted, immediately successful and as more and more connecting canals contributed their quotas of new traffic to the 'cross', dividends increased and the value of canal shares rocketed. Inspired by this success more and more new canal schemes were promoted all over the country, the process reaching its peak in an orgy of speculating in 1792 which is comparable with the great 'railway mania' of 1845.

This, very briefly, is the historical background to the story of the Ellesmere Canal. None of the canals of the cross or their connections touched the county of Shropshire; the great trade route of the shire from Bridgnorth to the Welsh March was still the river Severn. It had been a great water highway since medieval times, up-river trows trading from Bewdley as far upstream as Welshpool or Pool Quay as it was called. But, as Brindley was never tired of pointing out, such navigation in the upper reaches of rivers lacks the certainty of canal transport, and traffic movement on the upper Severn was often brought to a stand for weeks at a time by winter floods or summer droughts.

The agents of canal promotion in the county were the Shropshire ironmasters and their first creation was that strange little system of tub-boat canals comprising the Donnington Wood, Ketley and Shropshire Canals which connected the ironworks and mines in the neighbourhood of the Wrekin with the Severn at Coalport, and which was completed in 1792. The prime instigator and engineer of this scheme was himself an ironmaster – William Reynolds, partner of Abraham Darby of the Coalbrookdale Ironworks. Reynolds' particular contribution to canal engineering was his adoption of the inclined plane as an alternative to locks for overcoming changes of level, and his Ketley Canal included the first of these planes ever to be built in England. It was by the aid of a similar plane that the small tub-boats – floating boxes 20 ft. long by 6 ft. wide by 3 ft. deep – were lowered 213 feet down to the level of the Severn at Coalport where their contents were transhipped into the Severn trows.

As a solution to a purely local transport problem Reynolds's little Shropshire Canal was perfectly successful, but it was never intended to be, and could never become, part of any trunk waterway. The nearest full-scale waterway to the Shropshire border in 1791 was the Chester Canal and its history up to that date was melancholy enough to provide an Awful Warning for local canal speculators. Its line stretched from Chester to Nantwich and it was not connected with any other waterway except the treacherous sand-barred estuary of the Dee. It had projected a branch to join its great and prosperous neighbour, the Trent & Mersey, at Middlewich, but the reaction of the latter's proprietors was typical of contemporary canal politics. They evidently suspected that instead of bringing additional traffic to the Trent & Mersey the Chester Canal would be more likely to draw it away, using their communication with the Dee as an alternative to the Trent & Mersey's Preston Brook–Runcorn outlet to the Mersey. So they succeeded in getting a clause inserted in the Chester Canal Company's Act of 1777 which forbade their Middlewich branch to approach within fifty yards of the Trent & Mersey. Needless to say the branch was not built and as a consequence the Chester Canal remained isolated, impoverished and prematurely derelict.

This sad example of hopes unfulfilled did not deter the three gentlemen who met together in the little Flintshire village of Overton-on-Dee on Midsummer Eve, 1791. They were John Kynaston-Powell of Hardwicke, the Member of Parliament for Salop County, W. Mostyn Owen, the Member for Montgomery and the Reverend John Robert Lloyd who acted as Clerk and took minutes. They had met to discuss a proposal for a canal which would link the Mersey with the Dee at Chester and the Severn at Shrewsbury, serve the prosperous agricultural district of Shropshire with coal, lime and other goods and tap the rich coal and iron district around Wrexham and Ruabon. As they then conceived it, the main Dee–Severn line of the canal would run to the east of the river Dee, thereby avoiding the difficult hill country along the Welsh March to the west of the river. Unfortunately, however, this route also missed the Wrexham and Ruabon districts. In an attempt to mitigate this it was resolved to employ a local man, John Duncombe, to survey the line of a branch canal, eighteen miles long, which would cross the Dee at Overton, climb to a 210-feet summit at Ruabon, skirt the Wynnstay Park wall to Plas Kynaston and Trevor and then follow the northern slopes of the Vale of Llangollen to the Irenant Slate Quarries near Valle Crucis.

Powell and Owen's next step was to advertise a public meeting which was held at the Royal Oak Hotel, Ellesmere on the last day of August. Here it was decided to request the veteran John Smeaton to recommend an experienced engineer to make a survey. Smeaton gave them three names to choose from: William Jessop of Newark, a Mr. Eastburn of Odiham and Samuel Weston of Oxford. The promoters did well to select Jessop for he was a most capable engineer whose name would be deservedly remembered more widely today had he not been a very modest and retiring man. By this time another local man, William Turner of Whitchurch, had been making a survey and plans, and the promoters had become divided into two hostile camps, one favouring the original eastern route and the other a more difficult western route which would bring Wrexham and Ruabon on to the main line of canal. Although it is not explicitly stated in the records it is fairly certain that Duncombe and Turner were responsible for the preliminary surveys of the eastern and western routes respectively.

After he had considered the levels and sections and had been over the ground himself during the early summer of 1792, William Jessop decided in favour of a western line which, had it ever materialised, would have been one of the most costly canals ever built in this country. After climbing from Chester up to Wrexham and Ruabon by a ladder of locks, the canal would pass through a tunnel no less than 4,600 yards long and emerge from its southern portal to cross the Dee by aqueduct at a low level where that river leaves the Vale of Llangollen at Pont Cysyllte.[1] There would then follow more difficult cutting, including a second tunnel nearly a mile long, before the valley of the Ceiriog was crossed by another aqueduct at Pontfaen. The rest of the line southwards to Shrewsbury was comparatively easy except for a third tunnel a quarter of a mile long near the Shropshire village of Weston Lullingfields.

Even the most optimistic among the promoters were daunted by the idea of Jessop's 'great tunnel' as they called it, and he was requested to make fresh surveys in order to avoid it. The result was a route approximating to that which would eventually be built, although even when the Act incorporating the Company of Proprietors of the Elles-

1. Pronounced Kerssulty.

mere Canal received the Royal Assent on 30 April 1793 the problem of how the canal should be carried over the two major obstacles in its path – the valleys of the rivers Dee and Ceiriog – had not yet been solved. This was the position when Telford was appointed to the staff of the new company at the end of September.

It had been decided by the Committee of Management at their meeting in August that the post of 'General Overlooker' should be advertised in the local Press, but in spite of what has been written in the past there is no evidence that Telford applied for it and the tone of his letters to Little suggests that he did not do so. He made formal application for the post at the Committee meeting on 23 September, but it seems clear that his name was first put forward to the Committee without his knowledge, by the Shropshire ironmasters, Abraham Darby and William Reynolds.

On 29 September Telford wrote to Little in great jubilation giving him the news of his appointment. 'It is', he wrote, 'the greatest Work, I believe, that is now in hand in this kingdom and will not be completed for many years to come. You will be surprised that I have not mentioned this to you before, but the fact is that I had no idea of any such thing until an application was made to me by some of the leading Gentlemen and I was appointed at their meeting tho' many others had made much interest for the place. . . .'

'This is a great and laborious undertaking, but the line which it opens is vast and noble and coming in this honourable way I thought it too great an opportunity to be neglected, especially as I have stipulated my right to carry on my Architectural profession . . . It will require great exertions but it is worthy of them all. There is a very great Aqueduct over the Dee, besides Bridges over several Rivers which cross the Line of March.'

Telford's appointment had still to be ratified at a meeting of the General Assembly of the canal proprietors held at Ellesmere at the end of October and here it met with considerable opposition. 'You will not be surprised', Telford wrote, 'that altho' this employment was offered to me, that there should be many who looked forward to it with anxious eyes and that they had endeavoured to raise a party at the general meeting.' Telford is referring here to John Duncombe and William Turner who had been concerned in the project from the beginning, had done all the preliminary survey work and would understandably resent the appointment of a stranger over their heads. William Turner's application for the post was strengthened by the support of William Jessop. Yet Telford prevailed for, besides Darby and Reynolds he had, as he put it, 'the decided support of John Wilkinson, king of the Iron Masters, who is in himself an host'.

As we might imagine from this, the atmosphere in which Telford took up his appointment was scarcely a happy one. In addition to personal rivalries and jealousies among the staff, there was a division of opinion among the proprietors, many of whom still favoured the easier eastern route and thought the bold decision to follow the western line was ill advised. No one had attempted to engineer a canal through such difficult country before. 'Besides the real labour that attends such a great public work,' wrote Telford, 'contentions, jealousies and prejudices are stationed like gloomy sentinels from one extremity of the line to the other. But, as I have heard my mother say that an honest man might look the Devil in the face without being afraid, we must just trudge on in the old way.'

It has been suggested that in ascribing the engineering of the Ellesmere Canal to Telford he has been given more credit than is properly due to him. This suggestion is based on the fact that in contemporary documents Telford is often referred to simply as 'General Agent', whereas both William Jessop and John Duncombe are described as 'Engineers'. This has somewhat naturally given rise to the belief that Telford's position was merely that of a business manager while engineering responsibility remained with Jessop and Duncombe. Because such doubts exist it will be as well at this point to quote the exact terms of Telford's appointment as recorded in the Managing Committee's minute book in order to establish beyond further question what his position was. He was appointed: 'General Agent, Surveyor, Engineer, Architect and Overlooker of the Canal and Clerk to this Committee and the Sub-Committees when appointed. To attend meetings, make reports, to superintend the cutting, forming and making of the Canal and taking up and seeing to the due observance of the Levels thereof, to make the Drawings and to submit such drawings to the consideration and correction of Mr. Wm. Jessop or the Person employed by the said Company for the time being as their Principal Engineer. To give instructions for contracts, to attend himself (or some confidential person by him employed) to the execution of all the Contracts relative to the making and completing the said Canal. To pay Contractors and workmen and to keep accounts. His engagement to extend to all Architectural and Engineering business, to the drawing, forming and directing the making of bridges, aqueducts, tunnels, locks, buildings, reservoirs, wharfs and other works.' For this he was to be paid a salary of £500 a year, out of which he was to pay his own clerk and assistant, and to find a surety of £5,000 for his 'faithful performance'. In December, however, the salary provision had altered. It was agreed that he should be paid £300 a year and that the Company should pay his assistants. In the early stages of the work, Telford was saddled with the unpleasant duty of collecting money from the shareholders (no easy task!) whenever it was decided to make a call on the shares. But this was a responsibility which he soon managed to delegate so that he could concentrate on engineering matters.

It could still be argued from this that Telford was in the position of a resident engineer working under Jessop so that it would not be consistent to give Telford the credit for the Ellesmere Canal without conceding the same credit to men like W. A. Provis or Alexander Easton who later worked as resident engineers under Telford's direction. There is an important distinction here, however, for whereas Provis and Easton were engaged to carry out Telford's designs, in this case it was Telford who originated the designs and submitted them to Jessop for his approval. So it is safe to say that the major share of the credit for the great works on the Ellesmere Canal is due to Telford, although it is quite true that inadequate tribute has been paid in the past to that modest man, William Jessop.

Although he had backed Turner against him, Jessop did not harbour any ill will against Telford on that score. Jessop had worked with Turner and knew nothing of Telford. But he seems quickly to have appreciated Telford's ability and the two men immediately settled down to work amicably together. It was the beginning of a friendly association based on mutual respect and admiration which was to last until Jessop's death in 1814. Jessop seems to have realised that when it came to masonry work he had little to teach Telford. Telford, on the other hand, knew that in the craft of canal cutting he was completely inexperienced and, as he afterwards acknowledged in his autobiography, he consulted Jessop freely on the subject of the earthworks.

So comprehensive are the terms of Telford's appointment that they promise enough work to keep not one but six men fully occupied. Yet even after his engagement had been confirmed he could still write to Little: 'I have reserved the right to carry on such of my architectural business as does not require my personal attendance, so that I shall retain all I wish for of that, which are the public Buildings and Houses of importance. The other parts of our business are better to be without; they give a great deal of unpleasant labour for very little profit; in short they are like the calls of a Country Surgeon. These I shall give up without reluctance, except what relates to Mr. Pulteney and Lady Bath and I have pleasure to say that they are not disposed to quit me.' Although Telford continued his work on his Shropshire and Severn bridges and saw through to completion his churches at Bridgnorth and Madeley, in fact he abandoned his architectural career at this point and if he did carry out any further work for Pulteney there is no evidence of it. This was the great turning-point of his life and looking back upon it from the distance of old age he saw that it was so, though he attributed it then to a more deliberate act of will on his part than he actually made at the time. Writing of the Canal Company's invitation in his autobiography he said: 'Feeling in myself a stronger disposition for executing works of importance and magnitude than for the details of house architecture, I did not hesitate to accept their offer and from that time directed my attention solely to Civil Engineering.'

When work began on the Ellesmere Canal both John Duncombe and William Turner worked with Telford as his Assistant Engineers and Surveyors. It cannot have been a very happy partnership, for although Duncombe seems to have accepted his subordinate position quite philosophically, Turner did so with very ill grace. In February of 1794, however, Telford gained a stout ally when he secured the appointment of Matthew Davidson as Inspector of Works.

The first contracts to be let, both to Samuel Weston & Son of Oxford, were for the Mersey–Dee section between Netherpool (now called by the Company Ellesmere Port) and Chester and for part of the Llanymynech branch which had been planned to connect at Carreghofa with the Montgomeryshire Canal to Welshpool and Newtown which received its Act in March 1794. By cutting first what became known as the Wirral Line the Company showed good sense, for the going across the flat Wirral peninsula was easy and there was a good potential traffic between Chester and the Mersey. The decision must also have kindled a spark of hope in the breasts of the unfortunate shareholders of the Chester Canal Navigation. The work proceeded so swiftly and smoothly that the Wirral Line was opened for traffic in 1795 and at once began to earn revenue which helped to defray the cost of the more difficult works to the south. It was a broad canal, that is to say its locks could pass boats of fourteen-feet beam, the original proposal having been to construct the whole canal to this gauge. Not only was there a heavy merchandise traffic on it, but two passenger packet boats had been built to Telford's design ready for the opening. These ran from Chester to the Port, where they connected with the Mersey Packets to Liverpool. The popularity of this service may be judged by the fact that the Company leased the two boats and the Ellesmere Canal Tavern which was their terminal station to one Samuel Ackerley for no less than £1,000 per annum for two years. As Ackerley later renewed the lease he was obviously not a loser.

Steady progress was also made with the Llanymynech branch between Carreghofa and Lockgate Bridge, where it joined the projected main line to Weston Wharf and the Severn. On the construction of the most difficult part of the main line between Chester and Ruabon and across the valleys of the Dee and Ceiriog, however, there had been no progress at all although plans were drawn up for the two great aqueducts in 1794.

It is impossible adequately to appreciate the magnitude of Telford's achievement when he bridged the Vale of Llangollen unless one has seen his aqueduct at Pont Cysyllte and has some knowledge of engineering technique at the time it was projected. James Brindley had, many years before, successfully carried his canals across rivers by aqueducts which stand to this day. But he did so at low level, supporting the immense weight of the waterway in its puddled clay bed upon brick arches so squat that they resemble a series of enlarged culverts carried through the canal bank. In the years following Brindley's death nothing notably more ambitious had been achieved. Canal engineers had faithfully followed their master's technique which was totally unsuited to structures of any great height and was therefore no solution to the problem of carrying a canal across a deep valley at high level. It was Telford, a newcomer to canal engineering, who solved it by rejecting all precedent, and Pont Cysyllte, his first major work, represents his magnificent answer. It consisted of carrying the canal in a cast-iron trough upon stone piers over a length of 1,000 feet and at a height of 127 feet above the river. Sir Walter Scott called it the greatest work of art he had ever seen, and even today its effect upon the beholder is breathtaking. To stand upon it and look down at the water crawling over the boulders of the river bed far below is to sympathise with George Borrow's Welsh guide who, as they crossed over, confessed that: 'It gives me the pendro, sir, to look down.'[1] Pont Cysyllte is indeed one of the great engineering achievements of all time, so it is proper to consider in some detail how it originated; especially as there has hitherto been some difference of opinion on the subject. The story takes us first to another Shropshire waterway.

On 3 June 1793, the Shrewsbury Canal Company was empowered to cut seventeen and a half miles of canal to connect the Ketley tub-boat canal at Trench with Shrewsbury at Castle Foregate. The object of this was to by-pass the uncertain and tortuous Severn navigation in the conveyance of coal and iron from the Wrekin district to the county town. There was no connection with the river at Shrewsbury, but the proprietors hoped that a junction would eventually be made with the Ellesmere Canal. Needless to say the Shropshire ironmasters were once again very much to the fore in this venture. Narrow locks eighty-one feet long were to be built to pass either canal boats or trains of four tub-boats, but at Trench the new waterway was to be linked to the Ketley Canal by another of William Reynolds's inclined plane lifts. Reynolds was named as joint engineer of the canal with William Clowes.

Long before the Shrewsbury Canal was completed, William Clowes died and at some time between November 1794 and March 1795 Telford was appointed in his stead. In March Telford imparted this news to Andrew Little. He said: 'Since I wrote you I have been appointed Engineer to another Canal called the Shrewsbury Canal. It reaches from the Town of Shrewsbury to the Coalleries in the neighbourhood of the Wrekin at a place called Ketley where there are likewise great Iron Works. ... There are several Locks, two Aqueducts and a Tunnel underground for about ½ a mile. I have just recommended an Iron Aqueduct for the most considerable; it is approved and will be executed under my direction upon a principle entirely new and which

1. *Wild Wales*, Chapter 12.

I am endeavouring to establish with regard to the application of iron.'

The aqueduct to which Telford refers here crosses the river Tern near the village of Longdon. It is a modest single-span structure, but it is without doubt the earliest iron aqueduct in the world, and it is commonly regarded as the prototype of Pont Cysyllte. Notwithstanding Telford's own words on the subject of this aqueduct, William Clowes, Thomas Eyton the Chairman of the Canal Committee and, more plausibly, William Reynolds, have all, at one time or another, been credited with the original idea of using an iron trough. If any of these ascriptions were correct it would mean, not only that Telford claimed the credit for the Tern aqueduct under false pretences, but that his conception of Pont Cysyllte was not truly original. Happily the facts tell a different story.

William Clowes died before the Tern aqueduct was completed and Telford reconstructed it, probably using the original somewhat clumsy stone abutments built under the direction of Clowes. But in fact the course of events on the Ellesmere Canal prove that no claim on behalf of Clowes, Eyton or Reynolds can stand because Telford had produced drawings for an iron aqueduct at Pont Cysyllte in March 1794, nearly twelve months before he was appointed engineer to the Shrewsbury Canal.

It was on 17 January 1794 that the first plans for an aqueduct at Pont Cysyllte were laid before the Managing Committee at Ellesmere. The minute book records: 'that the plan of the Aqueduct at Pontcysyllte with three arches as prepared by Mr. William Turner of Whitchurch, Architect, shall be adopted by this Committee with such alterations as Mr. Jessop shall communicate to Mr. Telford, and that Mr. Telford do prepare a specification and proper sections and working drawings to enable workmen to give in estimates.'

Now this reference to a three-arch aqueduct makes it clear that Turner had designed a small structure of conventional pattern which would have crossed the Dee at low level. This meant carrying the canal down into the valley by locks at each side with all the additional expense and water-supply problems which this would involve. Nevertheless, the Committee were obviously prepared to accept the plan. No doubt they felt that the problem had been argued over too long and were desperately anxious to make a start. Telford, however, felt most dissatisfied with the scheme, but it was obviously no good his damning it unless he could produce a better idea. So he did some very hard thinking and it was undoubtedly during the fortnight following this meeting that the seed of his mighty iron aqueduct began to grow in his mind. What he told the Committee when they met again on the last day of January we do not know, but at his urgent request they consented to postpone advertising specifications of Turner's aqueduct for tender and to grant Telford the sum of £100 towards the cost of preparing detail plans of an alternative design. For William Turner, soured by his defeat at the General Assembly and jealous of Telford's success and growing authority, this was the last straw and he left the Company.

In March Telford paid a hurried visit to London to lay his plans before William Jessop, and having obtained his approval they were placed before the Committee on the last day of the month. They were passed and ordered to be advertised for tender in the Liverpool, Manchester, Birmingham and London papers. The most favourable tender came from James Varley of Colne, Lancashire, and in May Telford was ordered to negotiate a contract with him.

There is evidence in a letter from Telford to Matthew Davidson that James Varley was working at Pont Cysyllte at the beginning of 1795 if not before, but if this was so he can have done no more than clear the site. For it is evident, and scarcely surprising, that the Managing Committee were suffering from icy cold feet as the implications of Telford's proposal sank deeper into their minds. The idea of boats floating across the Vale of Llangollen at a height of 120 feet would strike them as so fantastic that they must have wondered uneasily whether their engineer was a genius or a lunatic. Their fears were preyed upon by William Turner who denounced Telford's plan as impracticable. Still smarting with resentment, he was trying to claim payment for the work involved in preparing his rejected plans and the Committee later complained about 'publicly misleading and unauthorised statements' which he had made in the local Press.

In the light of this situation in the spring of 1795 with all its doubts and head-shakings, the true significance of the Tern aqueduct on the Shrewsbury Canal and of Telford's remarks to Little in his March letter become clear. As soon as he had been appointed engineer to the Shrewsbury Canal and had surveyed the uncompleted line, Telford realised that at Longdon there was a heaven-sent opportunity to demonstrate upon a small scale his Pont Cysyllte design and so convince the sceptics that the use of a cast-iron trough was a practicable proposition. When he refers in his letter to '... a principle entirely new ... which I am endeavouring to establish', he is obviously alluding to his efforts to obtain whole-hearted support for his Pont Cysyllte plans. The Longdon aqueduct is not, therefore, the forerunner of Pont Cysyllte although it was completed first. It was merely the experimental application of a great design which had already been conceived when Telford's connection with the Shrewsbury Canal began.

It was fortunate for Telford that whatever misgivings the Managing Committee might harbour he had succeeded in convincing William Jessop that his plan was feasible, for without Jessop's approval it could never have gone forward. Jessop did, however, suggest modifications to Telford's original design which were adopted. This is revealed in a letter from Jessop to Telford dated 26 June 1795 which fortunate chance preserved in a file of Matthew Davidson's correspondence. 'In looking forward to the time', wrote Jessop, 'when we shall be laying the Iron Trough on the Piers I foresee some difficulties that appear to me formidable. In the first Place I see the men giddy and terrified in laying stones with such an immense depth underneath them with only a space 6 feet wide and 10 feet long to stand upon, and the same want of room will hardly allow space for the Beams and scaffolding while the Iron work is putting together.

'I therefore think in the first, that in order to reduce the weight of the Iron or parts of it it will be better to have the openings narrow by adding another Pier so as to have 8 openings of 52 feet from Centre to Centre instead of 60 ft.

'In the next place I would have the Piers 7 feet wide at the top instead of 6 feet, and make them about 2 feet more in the other dimensions. I hope you will not have proceeded with the Foundations till you received this. And I begin to think it may be better to have the canal 9 feet wide and 5 feet deep as was first intended.'

Jessop's reference to only eight openings is at first puzzling, because in fact Pont Cysyllte has no less than nineteen spans which, at the figure of fifty-three feet eventually adopted make a total length of 1,007 feet. The most likely explanation is that Telford had planned his aqueduct to extend only from the steep northern slope of the valley to the southern bank of the Dee as this distance is exactly accounted for by the eight loftiest spans. No doubt he planned to cross the gently

sloping meadows south of the river either by prolonging the embankment on that side or by building masonry approach spans. The reason
why this plan was changed is easy to understand. As soon as the cast-
iron trough had proved itself at Longdon it must have been obvious
that to extend it on this south side of the valley would be by far the
quickest, cheapest and most satisfactory method. As it was, the southern
approach embankment when completed reached a height of no less than
ninety-seven feet at its tip, being by far the greatest earthwork ever
raised in Britain at that time.

In one other important respect the completed aqueduct would differ
from the figures quoted by Jessop. The width of nine feet given for the
waterway in the iron trough shows that it was intended to be the same
as that used at Longdon. There the towing path is carried beside the
trough on cast-iron brackets, but Telford evidently decided that to
adopt the same plan over the much greater length of Pont Cysyllte
would not be satisfactory. In so long a trough only nine feet wide the
resistance to the passage of a loaded canal boat of the standard beam
of seven feet would be so great that the water would pile up in front of
the boat and overflow the trough. So Telford increased the width of
the Pont Cysyllte trough to 11 feet 10 inches and cantilevered the
towpath over the water on the east side of it.

Jessop having formally approved Telford's plans, the foundation
stone of the first pier was laid by Richard Myddelton of Chirk Castle
on the 25th of July 1795. Matthew Davidson was instructed to superintend the work on behalf of the Company and in December the Committee ordered a cottage to be built for him at the north end of the
aqueduct. Sir Alexander Gibb maintains that work on the piers was not
begun until at least fifteen months later. He bases this belief solely on a
letter which Telford wrote to Davidson on 7 October 1796 covering a
drawing of the aqueduct. In this he says: 'I have left the foundations to
be put in on the spot and according to their depth, the number of the
Piers are likewise undetermined.' It is practically certain, however, that
Telford is here referring to the southward extension of the aqueduct
from the river bank to the approach embankment. The 1795 date of
commencement is unlikely to be wrong as it is cast upon an iron com-
memorative tablet let into the base of the great pier immediately south
of the river which almost certainly marks the scene of the ceremony.
The masonry work was obviously well advanced by August 1797 or

Telford could never have informed Little then that the aqueduct was: 'already reckoned amongst the Wonders of Wales, for your old acquaintance [Davidson] now thinks nothing of having three Carriages at his Door at a time.'

James Varley's contract was for the piers only. William Davies, a local contractor, was responsible for the great south embankment, while it was not until 1802 that the ironwork was contracted for. With the object of relieving the foundations from unnecessary weight, Telford had designed the piers to be hollow from a height of seventy feet upwards. The external walls of this hollow portion were to be only two feet thick but it was strengthened by internal cross walls. This established an important new principle of construction which Telford employed on all his subsequent bridge works. It called for the highest degree of skill in masonry work and it soon became clear to Telford that Varley was not man enough to complete such a task unaided. The Company had expended a mere £2,000 on Varley's contract when Telford called to his rescue his 'treasure of Talents' John Simpson of Shrewsbury. Varley and Simpson continued to work together on a partnership basis until 1800 when the former disappeared from the scene and Simpson contracted to complete the piers alone. He was assisted, however, by another highly skilled mason named John Wilson of Dalston, Cumberland. Like Davidson and Simpson, Wilson was to be associated with Telford's works for the rest of his life.

At about the same time as the building of the Pont Cysyllte piers was begun, work on the section southwards through Chirk was started. This included two tunnels at Whitehouses and Chirk and the aqueduct over the Ceiriog. In the design of these tunnels Telford again departed from precedent, for he carried the towing path through them. Hitherto, canal tunnels had been built without towing paths which means that the boats had to be shafted or 'legged' by boatmen who lay on their sides and pushed with their feet against the tunnel walls. This was a slow, dangerous and exhausting job, so that many a boatman must have blessed Telford even though, on the Ellesmere Canal, the width of the path restricted the tunnel channel to one-way traffic. As the longer tunnel, at Chirk, was only a quarter of a mile, however, this was not a serious disadvantage.

The aqueduct at Chirk has always been somewhat overshadowed by the proximity of its greater neighbour, but on any other canal it would rank as the major work. It consists of ten spans of forty feet each carrying the canal seventy feet above the waters of the Ceiriog. Here again, as at Pont Cysyllte, both Telford and Jessop realised that the orthodox method of carrying the canal in its puddled bed would impose an impossible load upon such lofty piers and there was a great deal of discussion between the two men as to the best form of construction to adopt. Telford finally designed a masonry structure orthodox in exterior form but within which cast-iron plates, flanged and bolted together, were used to form a continuous metal bed for the waterway. At the sides, these plates were securely keyed in ashlar masonry backed by brickwork set in Parker's cement. In this way the iron bottom not only formed a watertight bed for the canal which was much lighter and more efficient than clay puddle, it also firmly tied the side walls of the aqueduct together and so eliminated all risk of the blowouts due to the lateral pressure of water and puddle which had bedevilled canal engineers in the past. For the tops of the piers and the spandrels of the arches Telford again used the hollow wall technique with interior cross walls. The foundation stone of Chirk aqueduct was laid on 17 June 1796 and it was completed in 1801.

The building of the Chirk aqueduct is of particular interest because it brings on to our stage the last, but by no means the least important, of those characters whose names from now on were to be closely linked with nearly all Telford's great undertakings. This was William Hazledine – 'Merlin' Hazledine as Telford nicknamed him. Like Telford, Hazledine was a Freemason and the two men had first met as fellow members of the Salopian Lodge in 1789. Hazledine was six years younger than Telford and came of an old Shropshire family. He had been apprenticed as a millwright under his uncle John who had a good business erecting corn mills and iron forges in the Wellington area. William's elder brother, John (with whom he is often confused) founded the famous ironworks at Bridgnorth where Rastrick was for some time a partner and where Richard Trevithick's first locomotives were built. William Hazledine had no connection with the Bridgnorth business, but became an ironmaster in his own right with foundries and forges at Shrewsbury (Coleham) and Plas Kynaston. The history of these two ironworks and of William Hazledine's connection with them has not yet been traced, but it is this writer's guess that at the time

Telford first met him he was working only the Coleham Foundry and that it was the Ellesmere Canal promotion which encouraged him to branch out on a bigger scale at Plas Kynaston.

If this was the case it was an astute move, for these new works were sited within a few hundred yards of the northern end of Pont Cysyllte upon ground now occupied by a part of the huge Monsanto Chemical Plant at Cefn Mawr. The last trace of Hazledine's ironworks did not disappear until 1949, while the short arm of canal which leads into the chemical works was built to serve Hazledine's needs.

On 10 February 1797, William Reynolds and John Wilkinson of Bersham Ironworks submitted a tender for the Pont Cysyllte ironwork but for some reason it was allowed to lie and no contract was placed. The most likely explanation is that at this time Hazledine was busily equipping his new ironworks and that Telford advised the Canal Committee to wait until he was in a position to submit a tender. This may sound like a case of 'jobs for the boys'; nevertheless in days when the transport of heavy castings presented such a tremendous problem, the existence of an ironworks so near the canal line would have been a great advantage to the Canal Company and no other ironmaster could hope to compete with Hazledine on a price delivered to site. We may assume that Plas Kynaston furnace was in blast by 1799 if not before, for in November of that year Hazledine tendered successfully for the Chirk aqueduct bed-plates. Two years and three months later, on 17 March 1802, he secured the contract for the Pont Cysyllte ironwork. This did not cover merely the supply of the trough sections and iron spans; it included their erection and assembly upon the dizzy height of the great stone piers which now paced across the Vale of Llangollen.

By the end of 1793 the speculative whirlwind of the brief 'canal mania' had blown itself out. It was succeeded first by an ominous calm and then by the icy grip of a financial slump which froze many a half-completed canal work. In such an economic climate it is not surprising that Telford's great aqueduct took ten years to build. Another result of financial stringency which really is remarkable was that in the passage of this decade the Ellesmere Canal main line abandoned both its original objectives and took a totally different course.

Apart from the financial difficulties which beset the Ellesmere Canal Company the chief agent in bringing about this change of plan was the physical difficulty of making the canal line from Pont Cysyllte to

Chester. Having vetoed Jessop's original plan with its 'great tunnel' the Managing Committee argued interminably over the problem of how best to carry their waterway through this difficult country, and survey after survey was made. It was a tantalising situation for here, in the collieries of Acrefair, Ponkey and New Hall and the iron furnaces of Bersham and Brymbo lay the chief sources of wealth and trade which the canal was designed to tap. The abandonment of the 'great tunnel' meant that the canal must climb by a flight of locks from the north end of Pont Cysyllte and then descend by an even greater flight through the valley of the river Alyn by Gresford into the Cheshire plain. Water supply to the short summit level between these two flights was the great problem. Reservoirs were planned in the narrow valleys which run up into the eastern slopes of Ruabon and Esclusham mountains and the cutting of a short feeder canal called the Froode Branch was actually begun. But Telford and Jessop were obviously unhappy about the situation. They knew that the success of a canal depends utterly on an adequate supply of water to its summit level because every boat which passes through will draw two locks full of water away from that summit, one as it ascends and another as it descends. Brindley always tried to ensure a long summit level which would itself act as a reservoir and his successors had tried to follow his example. But here Telford and Jessop were faced with the prospect of a short summit level doubtfully supplied.

The first question to be pursued was whether some device more economical in the use of water than the lock could be used for raising and lowering the boats. This was an idea which was exercising the minds of many aspiring inventors at this time and the canal mania fostered as lavish a crop of weird and wonderful devices as did the railway mania fifty-three years later. Among them was a vertical canal lift patented by two gentlemen of Ruabon by name Edward Rowland and Exuperius Pickering. This consised of a trough having watertight end doors into which a boat could be floated. This was exactly counterbalanced by means of what the inventors called a 'diving chest' which was attached to the bottom of the trough by means of pillars whose length depended on the height of the lift. This 'diving chest' was in effect a watertight wooden box, of approximately the same dimensions as the trough above, which rose and fell in a well sunk below the lower level of canal to a depth equal to the lift required. It was the buoyancy of this chest in the well which counterbalanced the boat trough so that the whole contraption could (in theory at least) easily be raised or lowered by means of a winch without, of course, any loss of water in the process. Evidently Messrs. Rowland and Pickering succeeded in convincing the Ellesmere Canal Company that their brain child would work, for Telford (with what misgivings?) was ordered to select a site for such a lift in December 1794. Evidently the lift was actually built and proved a dismal failure, for after Jessop and John Rennie had examined it in 1800 the Company agreed to pay the disillusioned inventors £200 in compensation for the £800 they had expended.[1]

Next, in 1797, the use of inclined planes of the Reynolds type was considered, and a Mr. Henry Williams of Ketley was sent for to look at the canal line in the Gresford neighbourhood with this end in view. This came to nothing and Telford and Duncombe were then dispatched to Marple to examine and report upon the Peak Forest railway[2] at that place. They reported favourably and thenceforward railways became the chief topic of discussion at the meetings of the Managing Committee. At one stage, so enamoured did the Committee become with this new idea, it was even suggested that instead of the canal a railway should be carried over Pont Cysyllte as far as Vron Cysyllte village where a temporary canal terminal wharf had been established pending completion of the aqueduct. Finally, all idea of connecting Pont Cysyllte directly with Chester either by canal or rail was abandoned. The only extension north of the terminal basin at Pont Cysyllte which materialised after years of debate was a tramway serving local collieries and ironworks which terminated at Ruabon Brook. It consisted of a double line of cast-iron rails weighing 40 lb. per yard length, William Hazledine of Plas Kynaston contracting to supply, lay and maintain them.

Meanwhile the southern end of the original main line had been the

1. The site of this experimental lift can only be conjectured as construction had not proceeded far by December 1794. Frankton locks would seem to be the most likely site because they are situated on one of the first sections to be built.

2. This was not the celebrated Cromford & High Peak Railway which did not receive its Act until May 1825. The reference is either to an earlier tramway from Bostol Limeworks to the Canal near Whaley Bridge, six miles, or to the temporary inclined tramway at Marple used while the sixteen locks there were under construction.

subject of similar debate and repeated survey until it had finally come to a stop at Weston Lullingfields where a terminal basin was built. Had matters rested thus it would have meant that the whole of the Ellesmere Canal and the Montgomeryshire Canal with which it connected would have been cut off from the rest of the canal system, including the Company's own Wirral Line and their optimistically named Ellesmere Port. The original project, however, had included a long branch canal from Frankton Junction, on the main line, through Ellesmere to Whitchurch. Work on this canal had proceeded smoothly and the Company saw in it the answer to their dilemma. By agreement between the two Companies the branch was extended north-eastwards from Whitchurch to join the Chester Canal at Hurleston near Nantwich. This extension was carried out by John Simpson in association with J. Fetcher, the Chester Canal Company's engineer who had previously made the survey with John Duncombe. The work was pushed forward with all possible speed and was completed in 1805.

So, at last, the Company achieved a through water route, albeit a very devious one, between Pont Cysyllte, Ellesmere, Chester and Ellesmere Port, while the long-suffering proprietors of the Chester Canal realised that their moribund dead-end waterway had suddenly become important.

Telford's services were now called upon to restore the Chester Canal to good repair. Its most obstinate defect had been the two locks near Beeston Castle, the chambers of which had been built upon quicksand with the result that they had given continual trouble owing to subsidence and consequent leakage. Telford's successful answer to this problem was to construct new lock chambers built entirely of cast-iron plates flanged and bolted together after the fashion of the bed of his aqueduct at Chirk and stayed to piles driven behind them. With every application of the new material his confidence was growing and with the unfailing co-operation of Hazledine he was becoming a master in the structual use of iron.

The abandonment of the northern section of the main line and the substitution for it of the Ruabon tramway, raised a new problem. This was how to provide an adequate supply of water for the canal from Pont Cysyllte downwards. The solution was to revive Duncombe's early survey of a canal line up the Vale of Llangollen to the Irenant slate quarries at Valle Crucis in the form of a navigable feeder to Llantisilio where it could be supplied naturally with water from the river Dee. It was also planned to raise the level of Bala Lake by means of a regulating weir and sluice so that in dry seasons more water for the canal could be released down the river. This project required the consent of the lake's owner, Sir Watkin Williams Wynn, and Telford arranged to meet him at Bala to explain the scheme and negotiate a settlement. Happily for the Canal Company, who would otherwise have been in the awkward position of owning a costly canal without any adequate means of filling it, Telford's diplomacy was successful, agreement was reached and the Act empowering the Company to make the feeder received the Royal Assent on 29 June 1804.

William Hazledine had promised to complete the ironwork on Pont Cysyllte in the summer of 1805 and the Company now made every effort to complete the feeder in time for the opening of the aqueduct. Desperate measures, including the selling of property, were taken to raise money to speed this work. While they scraped the bottom of the financial barrel in this way one wonders whether any of the Managing Committee members recalled with a twinge of conscience that bill of £102 2s. for wine supplied at their meetings which they had airily ordered their Clerk, Potts, to settle just three years before. Possibly the finances of the Ellesmere Canal Company improved somewhat when Telford's new canal office building at Ellesmere (now known as Beech House) was completed in 1806 and the Committee bade farewell to the convivial surroundings of the Royal Oak where they had met regularly for the previous fifteen years.

To cut six miles of canal along the steep northern slopes of the Vale of Llangollen, much of it through hard rock, was no light task and it very soon became obvious that to complete it in a mere twelve months would be quite impossible. It was not in fact finished until early in 1808 – the last section of the whole system to be cut. The fact that notwithstanding this delay the Company contrived to open their aqueduct according to schedule in 1805 poses a question to which the Company's records supply no answer. Most probably the Tref-y-Nant brook which rises in Ruabon mountain north of the Trevor rocks was diverted as a temporary, but inadequate, feeder. Such expedients must previously have been adopted further down the line to enable the canal to be opened to its temporary terminus at Vron Cysyllte basin.

The completion of his great aqueduct marked the end of Telford's service as agent, engineer and architect to the Ellesmere Canal Company and he handed over his office to two subordinates, Thomas Stanton who succeded him as agent and Thomas Denson who became resident engineer. But the Company retained him as their consultant, and until his death he continued to visit the canal twice a year to examine and report on the state of the works.

The ceremonial opening of Pont Cysyllte on 26 November 1805 must have been a proud moment for Telford and a scene which he would never forget. It was a fine, calm afternoon when the procession of six boats swung away from Vron Cysyllte wharf, moved slowly along Davies's great approach embankment and then sailed out on to the high iron trough. The first two boats were occupied by members of the Managing Committee and their families, the next contained the band of the Shropshire Volunteers playing lustily, while the fourth was filled by the engineers, their families and staff. Bringing up the rear were two empty boats destined to carry the first loads of coal across the Dee. As this procession moved out over the aqueduct a mighty cheer went up from the thousands of spectators who crowded the riverside meadows or watched from the hilltops, and the Vale of Llangollen echoed and re-echoed the thunder of cannon as a detachment of the Royal Artillery Company fired a fifteen-round salute. When the northern basin was reached there was more cheering as the first trams of coal came rumbling down the new railway to be shot into the two waiting boats.

By the time the procession turned to recross the aqueduct and cannons crashed out again the short day was nearly done; the sun had set behind the dark ruin of Glendower's hilltop fortress of Dinas Bran but as the November dusk thickened in the deep valley that wizard of the new world, 'Merlin' Hazledine, coloured the scene with his magic as the false sunset glow of his furnaces flared and flickered skywards from Cefn Mawr. As the boats glided along, their occupants gazed down in awe from that vertiginous height upon the river rushing over its boulders, upon meadows veiled in rising mist and upon the pale moons of upturned faces far below. To those standing by the river gazing skyward at the boats as they moved slowly along their iron channel, dwarfed by the tall perspective of Telford's towering piers, the wonder of it was almost past belief. This was Pont Cysyllte, the magical stream in the sky. Yet Telford's magic was of a very solid and material kind for, contrary to what some of those watchers may have supposed, it stands to this day sound and perfect in every joint and mortice. But the source of the greatest pride to Telford was this. In all the ten years of labour that went to the building of Pont Cysyllte, his care and foresight was such that he lost only one life.

Although the Ellesmere Canal never reached its original objectives and had a chequered financial history, as a feat of engineering it was a major triumph for Telford. He had entered the service of the Company as an ambitious young architect of no more than local repute; he left it a middle-aged civil engineer with a national reputation which placed him at the very head of his new profession. He had, moreover, while the canal was building, gathered about him with unerring judgement a small select company of men, William Hazledine, Matthew Davidson, John Simpson and John Wilson, each of whom was a past-master at his trade and whose combined skill was unequalled in England. When their work in Shropshire was done these men followed Telford to the scene of his next great task – the Highlands of Scotland, where he planned to drive a ship canal through the Great Glen.

5

ROADS TO THE ISLES

TELFORD'S attention was by no means exclusively devoted to the affairs of the Ellesmere Canal during the twelve years which elapsed between his appointment to the Company and the completion of Pont Cysyllte in 1805. Although he did not exercise the right of independence he had reserved to follow his old profession as an architect, he used that right increasingly as the years went by to pursue elsewhere his new profession as an engineer. During this period William Jessop's connection with the Canal Company seems to have become more tenuous as Telford became more self-reliant until, in fact if not in official title, Telford became engineer to the Company with Matthew Davidson playing the part of his resident assistant. His first journeys out of Shropshire were undertaken mainly on behalf of the Company, to Chester, to Liverpool and to London on parliamentary business or to confer with Jessop. But his influential position and the business connections which he formed as the Canal Company's agent soon bore fruit in the form of new commissions which entailed more and more travelling away from Shropshire. 'There is scarce a week', he told Little, 'but what I have to attend some public meeting.' And again: 'I am toss'd about like a Tennis Ball; the other day I was in London, since that I have to be in Bristol. Such is my fortune, and to tell you a bit of a secret, I truly believe that it suits my disposition.' One of these letters to Little, dated 26 February 1799, was written from the Salopian Coffee House, Charing Cross, which was to be his *pied à terre* in London for many years to come.

Notwithstanding the ever-increasing weight of his engineering responsibilities and the incessant coming and going which they entailed, he still pursued as ardently as ever that intense intellectual and imaginative private life which for him seems to have been a substitute for a wife, a family and a settled home. He was still writing poetry: 'It is to me something like what a fiddle is to others ... I apply to it in order to relieve my mind after being much fatigued with close attention to business.' The death of Robert Burns in 1796 moved him deeply and his poem to the poet's memory is one of his most successful efforts. Telford believed that Burns's removal from his farm to a post in the Department of Excise, where he aroused the hostility of officialdom, was a disaster for the poet, and his lines on this subject have a fine denunciatory ring about them:

> The Muses shall that fatal hour
>   To Lethe's streams consign,
> Which gave the little slaves of pow'r,
>   To scoff at worth like thine.

> But thy fair name shall rise and spread,
>   Thy name be dear to all,
> When down to their oblivious bed,
>   Official insects fall.

He was still reading widely. Stewart's *Philosophy of the Human Mind*, Adam Smith's *Wealth of Nations* and the poems of Goethe and James Thomson of *The Seasons* are examples mentioned in letters to Little which also reveal how fondly he still cherished his memories of Eskdale and of his friends at Langholm whom he continued to help financially and in numerous other ways. His mother was now old and failing and he contrived to pay one hurried visit to Eskdale to see her for the last time in the autumn of 1794. He did not look forward to this visit, for in a letter written to Little shortly before he set out from Shropshire he said: 'I am rather distressed at the thoughts of coming down as I must see a kind parent in the last stage of decay on whom I can only bestow an affectionate look and leave her. Her mind will not be much consoled by this parting, and the impression left upon mine will be more lasting than pleasant.'

As if all this public and private activity were not more than enough for one man, Telford also contrived to produce during these years a *Treatise on Mills* for the Board of Agriculture and to contribute a section on canals to Plymley's *General View of the Agriculture of Shropshire*. The original manuscript of the former is now in the Library of the Institution of Civil Engineers. It is a bulky volume beautifully illustrated by over thirty line and wash drawings, some by William Jones and others by Telford himself.

Although Telford's political judgements were unsound, some of the comments which his journeyings prompted were remarkably acute. Of Bristol and Liverpool he wrote: 'Bristol is sinking in its commercial importance, it is not well situated and its Merchants are rich and indolent – it will dwindle away. Liverpool is young, vigorous and well situated, it has, besides, taken *Root* in the Country by means of the Canals. Another port will rise somewhere in the Severn, but Liverpool will become of the first commercial importance.' Notwithstanding I. K. Brunel's belated efforts to inject fresh life into the Port of Bristol, this prophecy was destined to be fulfilled to the letter, anticipating as it does the latter-day growth of the new Severn port of Avonmouth.

Telford always remained at heart a countryman. When he rode or drove through England his eyes were always on the fields and his letters are seldom without some reference to weather or season or to the state of the crops. So it was that, although as a maker of canals and roads he became one of the chief agents of the Industrial Revolution, he foresaw very clearly the folly of developing industry at the expense of agriculture. In November 1795 when Europe was torn by the French war and Telford was at Derby on what he called 'one of his long circuits', he wrote: 'I am afraid that it is now too evident that of late years the Commercial and Manufacturing interests have increased too far beyond that of the Agricultural, so that the produce of the Land has not borne any just proportion to that of population and Commercial Wealth. While other Countries produced more than they could consume and while we were at peace with them the evil which was ripening at home was not perceived; but now that War has desolated all the Corn Countries of Europe and we must look to our own Stock Yards and potato Bings for supply the film is removed from our Eyes! . . .' So the words of Telford the engineer echoed those of that other much-travelled countryman, his great contemporary William Cobbett, but their warnings went unheeded.

New water-supply schemes for Liverpool (1799–1802) and Glasgow (1806), the improvement of the Severn Navigation (1800), the improvement of London's docks and the rebuilding of London Bridge (1797–1803), these were only a few of the schemes on which Telford was consulted during these busy years. Some of them may be referred to in subsequent chapters, for his life from this time forward became so full that any attempt to record his activities in strict chronological sequence would make confusing and tiresome reading.

In 1796, Telford was asked by the British Fisheries Society, of which Sir William Pulteney was Governor, to make experiments with Parker's cement and report on its suitability for the underwater masonry of harbour jetties and breakwaters. This must have seemed a very small assignment, yet out of it there grew the greatest undertaking of his life. This was no less than the opening up of land and sea communications throughout the Highlands and islands of Scotland. In order to understand how this came about and to appreciate the size of the work which Telford undertook it is necessary to know something of the tragic history of the Highlands in the fifty years which had elapsed since the spirit of the Clans was broken on Culloden Moor.

The ruthless policy of 'Butcher' Cumberland destroyed for ever the Clan system which had shaped the Highlander's way of life for centuries and created nothing to take its place. This life of the clansman, primitive and savage as that of the Border reivers and recognising no law or loyalty outside the clan, had already become an anachronism in the Britain of 1745. But it had been bred in the bone and its destruction by Wade's army left the once proud people of the glens as lost and as hopeless as the 'displaced persons' of Europe in our own unhappy century. The Government of the day raised no hand to help them. It was concerned only to break the claymore, not to forge it into a ploughshare. It is literally true that the plough did not exist in the Highlands in the eighteenth century; there was only the caschrom, a primitive form of breast plough whose ancestry is lost in the mists of Celtic prehistory when pastoral man first learnt to scratch the soil for food.

Apart from sword and fire the most important instrument in the breaking of clan loyalties was the confiscation of the estates of the chieftains who had supported Charles Stuart. This proved so effective that the last act of the drama was in many ways the most tragic of all. For when these forfeited estates were ultimately handed back in 1787, the heirs of the old chieftains had so far forgotten the paternal tradition of their fathers that, with a few honourable exceptions, they betrayed their people by evicting them and turning their crofts into sheepwalks, while the people, not without reason, no longer trusted them. After his first tour of the Highlands, Telford put the matter simply when he wrote: 'The chieftains may fight each other, the de'el a high-land man will stir for them. The Lairds have transfered their affections from the people to flocks of Sheep and the people have lost their veneration for the Lairds. . . . It is not a pleasant change. There was great happiness in the patriarchal state of Clanship; they are now hastening into the opposite extreme; it is quite wrong. . . .'

All that Wade's armies left behind them when the Rebellion had been crushed, apart from fire-blackened ruins, was the system of forts and military roads. The latter totalled some 800 miles. One joined the forts in the Great Glen with Inverness. A second linked this road with the Lowlands via the Pass of Glencoe, Tyndrum, and Loch Lomond side. A third ran from Fort Augustus through the Grampians to Blair Atholl; a fourth connected Fort George with Cupar-in-Angus via Badenoch and Braemar. But these roads had been constructed by military conquerors for strategic reasons. For the movement of troops they served their purpose but they benefited the broken clansmen scarcely at all. So, although they were the only metalled roads in the Highlands, a great part of them had fallen into disuse and disrepair by the end of the century and such little trade and intercourse as there was still moved along the ancient cattle tracks. Telford thought little of Wade's roads. Even for the purpose for which they were designed he considered that their routes had been badly surveyed. Their gradients were often needlessly steep, which was one of the reasons why the local people forsook them for the old tracks. Wade's Fort Augustus–Blair Atholl road, for example, climbs to a height of 2,500 feet at the Pass of Corrieyarrach and has consequently served no useful purpose from that day to this.

The result of the Highland tragedy was emigration upon such a scale that towards the end of the century it was realised that if no concerted attempt was made to check it the Western Highlands and islands would soon be completely depopulated. As a result of growing concern at this state of affairs, the Highland Society was formed in 1784 and a small group of M.P.s led by Henry Beaufoy, the member for Great Yar-mouth, and George Dempster of Skibo, representing Perth Burghs, began to press the urgent need for action upon the Government. In 1786, Beaufoy and Dempster, with Pulteney, the Duke of Argyll and others, founded the British Fisheries Society as a Joint Stock Company with the object of developing the Highland fishing industry by constructing a series of small fishing ports. This venture got away to a flying start, but after it had built its first two ports at Tobermory and Ullapool the Company's finances were crippled by the financial slump which followed the canal mania year of 1792. Moreover, the founders of the Company had realised from this small effort that the magnitude of the work of rehabilitation needed in the Highlands was far beyond their resources even if the financial climate improved. For it had become obvious to them that to build fishing ports was of very little use in a country which possessed no roads. If their object was to be achieved, ports and roads must be constructed simultaneously and if such a large-scale plan was to be put in hand Government assistance was essential.

Dempster, particularly, was tireless in pressing this case on the Government. He eventually obtained a promise from Pitt of Government help in organising a comprehensive survey of Highland roads, and in 1799 his brief was strengthened by a report from the Superintendent of Military Roads in the Highlands, Colonel Anstruther, which stressed the appalling state of communications. That state had become the more striking by contrast with the developments which had taken place in the Lowlands of Scotland since the mid-century. There, as in England, the Industrial Revolution was taking its course. Coal mines, ironworks and other industries were springing up, cities and towns were growing fast and in 1790 Robert Whitworth completed the canal from the Forth to the Clyde which Smeaton had begun. Yet north of the Highland line was a wilderness unchanged since the '45 and very difficult of access except by sea to such harbours as existed. It was impossible to travel by land from Edinburgh to Inverness without having to cross a river by ferry. The central route was broken by the Tay crossing at Dunkeld and the eastern by the Spey at Fochabers. When these rivers were in spate such ferry crossings became either extremely hazardous or impossible. When the Findhorn was in flood no one dare attempt the passage at Forres and weary travellers for Inverness must needs trek twenty-eight miles up Strath Dearn to General Wade's bridge over the river at Dulsie. North of Inverness the position was even worse. Drovers bringing their cattle southward had to swim them across the unbridged waters of the Conon and the Beauly and it was said that there was scarcely a wheeled cart to be found in the whole of Caithness. As for Sutherland, the only track into the county from the outside world lay over a rocky beach and was only passable at low water.

At long last, in the summer of 1801, the Government took action. On 27 July Telford, as engineer to the British Fisheries Society, received a letter from Nicholas Vansittart, Joint Secretary to the Treasury, instructing him to proceed with a survey and report on the subject of Highland communications. Detailed instructions, Vansittart told him, would follow, but Telford did not wait for them; by the time they arrived in September he had already completed most of his Highland journey. Here was an opportunity to display his powers and at the same time to serve his countrymen which he seized with both hands. Yet his precipitate departure was not solely due to his eagerness but also to his anxiety to complete his survey before winter gripped the Highlands and travel became impossible. As it was he did not get back to Shrewsbury until 18 November, although the weather was extraordinarily kind to him.

'Never was there a Season more favourable for making Surveys,' he told Little soon after his return, 'I passed along the Western and Central Highlands, from thence to the extremity of the Island, and returned along the Eastern Coast to Edinburgh and scarcely saw a Cloud upon the Mountain's top.

'It would require a Volume to specify anything like the particulars of this journey. I shall therefore only say that every part of my Survey exceeded my expectations and I did not leave anything unaccomplished. ...' If only Telford had been able to set down some account of his travels the result might have been as illuminating as Boswell's earlier *Tour of the Hebrides*. But he drove himself so hard that he scarcely had time to write even the briefest of letters. From Peterhead on his way south he did dash off a letter to Dr. James Currie, one of his Liverpool friends, which is slightly more explicit. In this he said: 'I have carried regular Surveys along the Rainy West through the middle of the tempestuous wilds of Lochaber, on each side of the habitation of the far famed Johnny Groats, around the shores of Cromarty, Inverness and Fort George, and likewise the Coast of Murray. The apprehension of the weather changing for the worse has prompted me to incessant and hard labour so that I am now almost lame and blind.'

In Edinburgh, Telford, his mission completed, did pause for a week's well-earned relaxation in the company of Professors Stewart,

Gregory, Playfair and Robison of Edinburgh University. These men, headed by the ageing Professor Robison,[1] the lifelong friend of James Watt and teacher of John Rennie, were to Edinburgh what the celebrated Lunar Society was to Birmingham and Telford described their intellectual companionship as 'all the heart of Man could wish'. Here, too, he found Colonel Dixon, a kinsman by marriage of the Pasleys of Craig, who was then carrying out great improvements in his native county of Dumfries. Telford had originally planned to visit Little in Eskdale on his return journey, but Dixon insisted upon bearing him off to his estate at Mount Annan where he made a further stay before posting directly back to Shrewsbury. There he found arrears of Ellesmere Canal business waiting for him which kept him busy 'almost day and night', as he put it, and as soon as this was done he set himself to work on the 'Charts, Plans, Estimates and Reports' of the Scottish Survey.

The importance which Telford attached to this work and the tremendous amount of labour and thought he put into it is revealed in his next letter to Little dated from London on 14 April 1802. 'I have been kept in continual motion from the Country to the Town, and from the Town to the Country,' he wrote. 'It cannot be hard to conjecture where my leading object has been and where my chief anxiety centred. Never when awake – and perhaps not always when asleep – have my Scotch Surveys been absent. The discussions respecting Scotland have been postponed and are now in waiting for them. My plans are completed, the Draft of my Report has been made out . . . and I hope to deliver in the whole in the course of a few days. I shall then be fully committed, and then, the L . . D have mercy upon me – or rather, may they be productive of good and that good benefit Scotland.'

His report was duly submitted and was so favourably received that he was at once ordered to continue his survey during the coming summer with more particular reference to road communications and to make a further report. 'I have this day arranged the outlines of my Scotch Survey for this Season with Mr. Vansittart,' he advised Little

---

1. Telford had first met Robison in 1796 when the latter was on a visit to John Wilkinson's Ironworks at Brymbo. In a letter to Davidson, Telford then described him as 'a douce old man and cantie – religious – and an aristocrat but candid and moderate'.

from London on 16 June, and within a very few days he was northward
bound once more.

One of Telford's wide terms of reference was to investigate the
causes of emigration from the Highlands, although one might think
that these would be obvious enough even to the most boneheaded
government official. In his comments under this head, Telford the
countryman came to the fore and displayed his radical good sense –
radical in the oldest and truest meaning of that word. His complaint
was the same as that made by Sir Thomas More at the time of the
Tudor enclosures in England: sheep were eating men in the Highlands.
Bluntly, he presented the alternatives to the Lords Commissioners of
the Treasury. If they chose to regard the Highlands with the cold eye
of the economist as a food-producing district of the British Empire
then the process of turning crofts into sheepwalks would continue and
so, inevitably, would emigration. The alternative was to create a better
way of life for the people of the Highlands by developing the fishing
industry and the agriculture of the crofters, the basis of which had
always been the cattle which the sheep were ousting. Mutton or Men –
the issue was as simple as that.

New life and hope could come to the Highlands, he believed, only
through the veins of improved communications by land and sea and
he went on to outline his grand scheme: new and improved harbours;
new bridges on the roads from the south to Inverness; a canal through
the Great Glen to unite the east and west coast fishing grounds; roads
northwards from Inverness into Caithness and Sutherland; roads
westwards from Fort William and the Great Glen to Arisaig, to
Knoidart, to Lochalsh and the Hebrides.

Telford's second report, which he issued in 1803, made a deep
impression on the Government, while his countrymen honoured him
with a Fellowship of the Royal Society of Edinburgh. Having delayed
so long, the Government now acted swiftly, for in July of this year
Acts were passed which set up two Commissions, one for the Caledonian Canal through the Great Glen which will be the subject of the
next chapter, and one for Highland Roads and Bridges. The latter body
was also given responsibility for Fishing Harbours and Ports. The
Canal Commissioners included representatives of both Houses, the
Roads and Bridges Commissioners consisted only of the Commons
nominees who were the same in each case and included Sir William

Pulteney. The two Commissions also shared the same staff, who formed a remarkably efficient and harmonious working team. They were: John Rickman, Secretary, James Hope, Law Agent and Thomas Telford, Engineer. Rickman became a close friend to Telford until his death.

The Highland Roads and Bridges Commissioners functioned in a fashion very similar to the Baronial Guarantee system under which railways were built in the more remote districts of western Ireland many years later. When a scheme had been approved and a survey and plan had been drawn up by Telford it was financed by the Commissioners and by local authorities or landowners in equal proportion. For the improvement of Scottish ports the Commissioners also had at their disposal a balance of £35,500 from the Forfeited Estates Fund.

The setting up of this administration marked the true beginning of an immense civil engineering undertaking which was carried on under Telford's supervision for the next eighteen years. In that time 920 miles of new roads were built in the Highlands and in addition 280 miles of military road were realigned and remade. This involved the building of over a thousand new bridges which varied in size from little stone culverts over mountain burns to great bridges of stone or iron which spanned Spey and Tay, Beauly and Dee. In addition to this there were the numerous harbour works on which Telford advised or which were carried out under his direction.

At mention of the name of Telford most people think at once of his suspension bridges over the Menai and Conway or, if they know more of engineering history, they will add to these his aqueduct at Pont Cysyllte. It is natural that this should be so because these three bridges were Telford's greatest individual works; they are embodiments of his engineering genius which imagination can readily grasp. It is quite impossible to evaluate in the same way a system of roads, bridges and harbours spread over a whole country, nor would the most exhaustive catalogue of the individual works involved help us towards an imaginative realisation. It usually happens in this way that if an engineer is remembered outside his profession it is by some *tour de force* which captures popular fancy while his record of solid achievement is forgotten. Judged in terms of the sheer magnitude of the work involved and of its historical importance there can be no doubt that Telford's work in the Highlands was the greatest achievement of his career.

Telford organised the road construction programme into six divisions, each with a superintendent in charge, and appointed one general superintendent as his resident deputy in the Highlands. His first choice for this vital office was John Duncombe but it proved mysteriously unfortunate. Obviously Duncombe must have distinguished himself as Telford's assistant on the Ellesmere Canal or the latter would never have nominated him for such an important and responsible post, yet for some reason Duncombe proved a miserable failure in the Highlands and for the five years during which he clung to his office he seems to have been carried along by the efforts of two assistants, Alexander Easton and John Mitchell. Telford gradually lost patience with him. 'He seems to be getting into his dotage,' he complained to Rickman in October 1809, 'there is no getting him to finish things in time.' The end came a few months later when Duncombe died in prison in Inverness. 'I am quite vexed about the old fool,' wrote Telford, again to Rickman, 'his dying will not be a matter of regret, but in a Jail at Inverness is shocking.'

What misdemeanour landed poor Duncombe in prison is a mystery, but this apart his tragedy would seem to have been that of an ageing man in a strange country struggling on but no longer able to satisfy Telford's exacting demands. His failure obviously exasperated his chief, but this cannot excuse such callous comment on a man who had once given him good service. For once it reveals Telford in a most unfavourable light; but the provocation must have been great, for such high-handed intolerance did not come naturally to him or he would not have found so many men to serve him so long and so faithfully.

Duncombe was succeeded by John Mitchell, who became a tower of strength. Mitchell was a working stonemason from Forres who could scarcely read or write but whose outstanding abilities Telford quickly recognised. A great, raw-boned man, Mitchell soon became known throughout the Highlands as 'Telford's Tartar'. He possessed great physical strength, immense energy and an inflexible will which drove him so hard that even his iron constitution could not stand the pace indefinitely. Every year for fourteen years Mitchell covered an average of 10,000 miles by gig, on horseback or on foot through that wild country in every sort of weather. He would arrive late and soaked to the skin at some lonely sheiling, sit dozing and steaming by the turf fire till dawn broke and then be on his way once more. In such a fashion did John Mitchell work himself to his death in September 1824, when he was succeeded by his son Joseph who had followed his father's trade as a stonemason and proved himself no less capable.

The six Highland districts and their superintendents were as follows:

(1) Argyll,   Robert Garrow.
(2) Badenoch,   George Macfarlane.
(3) Lochaber,   Daniel McInnes.
(4) Skye,   James Smith.
(5) Ross-shire,   Robert Murray.
(6) Caithness &
       Sutherland, Thomas Spence.

The first road to be constructed was that from Fort William to Arisaig with the object of opening a line of communication with the Western Isles. This followed the north shore of Loch Eil, crossed the head of Loch Shiel at Glenfinnan, where Prince Charles had raised his standard, and thence passed by Kinloch Ailort to its end on the western seaboard. It was a difficult route and it is significant that when, many years later, the railway builders came to drive the extension of the North British Railway through to Mallaig they followed very closely the line of Telford's road.

North of this Arisaig route, two other roads were driven westwards from the Great Glen to the Sound of Sleat, the first from Loch Oich through Glen Garry to Loch Hourn and the second from Loch Ness through Glen Morriston and Glen Shiel to Kyle of Lochalsh, where the short Kyleakin ferry crossing linked it to the new roads on Skye. On the eastern coast, Dingwall became a new centre from which Telford's roads struck westwards through Strath Bran to Lochs Carron and Torridon and northwards to his great iron bridge at Bonar over Dornoch Firth. Here the northern road forked, one branch driving due north through the heart of Sutherland to Tongue and the other bearing north-eastwards through Caithness to Wick and Thurso. Telford carried this last road across the head of Loch Fleet by a great embankment which subsequently gave its name to the neighbouring Mound station on the Highland Railway and which later carried that Company's Dornoch branch line across the water.

The prodigious labour and the sheer physical hardship involved in driving such roads defies imagination, although those who are familiar with the far north may have some conception of it. The district inspectors drove themselves as hard as Mitchell. Robert Garrow, the Argyll inspector, reckoned that he covered 5,000 miles in a year on foot. Until 1824 the road makers laboured for weeks at a time on bleak moorlands or by stony, stormswept lochsides with no shelter but tents or improvised turf-walled huts. After that year, at Mitchell's suggestion, they were equipped with large four-wheeled caravans containing cooking stoves which must have been the ancestors of the similar vehicles used to this day by road-rolling contractors. On the Tongue road, lime for mortar had to be carried for distances of twenty miles and more on the backs of pack horses. In Skye and parts of the north-west no suitable stone for bridge building could be found; all had to be brought by sea to the nearest landing. 'It was scarcely possible', wrote Telford in his autobiography, 'to procure stones for covering 2 ft drains.'

Of the innumerable bridges which were built in the course of this vast construction programme it will be sufficient to list ten of the most notable ones, the figures in brackets after each indicating the spans in feet:

Ballater, river Dee, Aberdeenshire, 1807–9 (34, 55, 60, 55, 34).
Bonar, Dornoch Firth, Sutherland, 1811–12 (50, 60, 150).
Conon, river Conon, Ross and Cromarty, 1806–9 (45, 55, 65, 55, 45).
Craigellachie, river Spey, Banffshire, 1812–15 (150, 15, 15, 15).
Dunkeld, river Tay, Perthshire, 1806–8 (20, 74, 84, 90, 84, 74, 20).
Fairness, river Findhorn, Nairn, 1814–17 (36, 55, 36).
Helmsdale, river Helmsdale, Sutherland, 1811–12 (70, 70).
Lovat, river Beauly, Inverness, 1811–14 (40, 50, 60, 50, 40).
Potarch, river Dee, Aberdeenshire, 1811–14 (65, 70, 65).
Wick, Water of Wick, Caithness, 1805–7 (48, 60, 48).[1]

In these, as in many other smaller bridges of stone, Telford made good use, in constructing the spandrels of the arches, of that hollow wall technique which he had first employed in the Ellesmere Canal aqueducts. This not only reduced the weight on the piers; the internal walls also strengthened the structure whereas the old method of rubble-

1. The Bonar bridge has been replaced. Conon Bridge was found too weak to withstand the heavy loads passing over it in connection with the building of the new Atomic Plant at Dounreay. A Bailey bridge has been superimposed on the original piers and abutments pending the construction of a new bridge. With these two exceptions all the bridges listed and many others remain in use today exactly as built by Telford.

filling merely added great weight without strength. As will be seen from this list, the largest of the Highland stone bridges was that over the Tay at Dunkeld, which was the key to the Grampian route to Inverness via the Killiecrankie and Drumochter passes. This is one of the finest of Telford's stone bridges.

The main spans of the Craigellachie and Bonar bridges were of cast iron and in their design, so graceful, so delicate in its nicety of proportion as to suggest an impossible fragility, Telford displayed his unequalled mastery in the use of this material. In each case the reason for the decision to use iron was the same: the breadth and depth of the channel to be crossed and its liability to very heavy flooding. In the case of Bonar the masonry work was carried out by John Simpson with a partner named John Cargill of Newcastle. The ironwork for both bridges was cast by William Hazledine at Plas Kynaston and delivered by canal and sea. When it was on site, Hazledine's foreman, William Stuttle, and two assistants were sent up to superintend assembly and erection. When Matthew Davidson followed Telford to Scotland, Stuttle had moved into his house at Pont Cysyllte. Although he was responsible for erecting all the bridges cast by Hazledine for Telford, he seems to have been ignored or forgotten by other writers on the subject. Yet he has his memorial, for his name is cast on the iron bridge at Bettws-y-Coed.

The building of Craigellachie went smoothly, but at Bonar Telford ran into great difficulty and for the first time in his life he was compelled to modify a design radically after construction had begun. Although in this case there were no tragic consequences, the reason was the same as that which forced Thomas Bouch to alter the design of his Tay Bridge – Telford had been falsely led to assume a rock foundation on both sides of the Firth and on this assumption had designed an iron bridge of two 150-foot spans. These were exactly the same as the single span at Craigellachie, the idea being that Hazledine could cast all three to the same patterns. It was found, however, that the rock base did not exist on the site selected for the south pier and the second iron span had to be abandoned.[1] Two masonry piers and an extended abutment took

1. It is highly probable that this redundant span was subsequently used by Telford for the Esk bridge near Carlisle. This had two iron spans of 105 ft. and one of 150 ft. As this Esk bridge was replaced in 1916, it is not possible to prove the point by comparison.

its place. Telford was unhappy about this alteration, which meant that of all his Highland bridges Bonar alone is asymmetrical. Nevertheless to every eye but his Bonar appeared a magnificent achievement. Robert Southey, in his diary, recorded the reaction of a native of Sutherland on seeing the bridge for the first time. 'As I went along the road by the side of the water,' he is quoted as saying, 'I could see no bridge; at last I came in sight of something like a spider's web in the air – if this be it, thought I, it will never do! But presently I came upon it, and oh, it is the finest thing that ever was made by God or man!'

If any local people had doubts about the strength of Bonar Bridge they were set at rest by two incidents which occurred soon after it had been opened. First an enormous mass of fir-tree logs embedded in pack ice floated down the river Oykell and dealt the bridge a tremendous blow. Shortly after this a schooner drifted into it and carried away two masts, but under both these assaults Telford's spider web of iron stood fast.

It was just after work had begun on Bonar Bridge that the value of Telford's work in the Highlands was demonstrated by a most appalling tragedy from which John Mitchell had a providential escape. Mitchell was spurring his horse westward from Dornoch in an effort to catch the boat at Meikle Ferry which was then the only means of crossing Dornoch Firth. When he reached a point of vantage overlooking the ferry he realised that he was too late and that the boat had just left. Then, to his horror, he saw the crowded ferry suddenly capsize in the swirling tidal waters. Of the 119 souls on board her, only six managed to struggle to the shore.

In the work of Highland harbour improvement, Telford's first lieutenant was John Gibb of Aberdeen, founder of a distinguished line of civil engineers. Gibb was also placed in charge of the Crinan Canal improvement scheme in 1817 on which £19,400 was expended under Telford's superintendence. It would be tedious to deal in detail with the harbour works which were carried out during these fruitful years. Moreover in some cases Telford extended the works of previous engineers, notably John Smeaton, and in others he worked to improvement plans which had previously been drawn up by Smeaton or Rennie and shelved by the ports concerned because of lack of capital. Hence in the case of harbours, unlike any other works, it is often very difficult to determine how much of the work carried out in these years represents Telford's original conception. It is sufficient to say, then, that the more important harbour improvement works completed under his direction were those at Aberdeen, Dundee, Peterhead, Banff, Frazerburgh, Fortrose, Cullen and Kirkwall, in descending order of magnitude.

When the roads and bridges programme had been to all intents and purposes completed the Government decided to round off their belated atonement for past crimes against the Highlands by building a number of new churches and manses in remote districts. A Commission was set up for the purpose in 1824 with Rickman as secretary and Telford was asked to submit designs. These could be called Telford's last purely architectural works, that is if his designs deserve the name, for they are austere enough to satisfy the dourest minister of the Wee Free. To do Telford justice, however, he was given little scope for the display of architectural graces, for the Government stipulated that church and manse together must not cost more than £1,500. This was a very tight figure when we remember the cost and difficulty of transporting building materials to such remote places. Joseph Mitchell and Robert Garrow were appointed to supervise this construction programme, Mitchell covering the northernmost district. In all, thirty-two new churches and forty manses were built at a cost of a little over £54,000. Telford churches are to be found on Iona, Ulva, Lewis and Islay, on Mull, Harris, North Uist and Skye, on Quarff in the Shetlands and North Ronaldshay in Orkney.

Telford's road work in Scotland was not solely confined to the Highlands. He supervised in addition the construction of 184 miles of new road and several notable bridges in the Lowlands. John Pollok was his superintendent for this Lowland district where the chief works were the greater part of the present trunk route from Carlisle to Glasgow with notable bridges at Hamilton, Birkwood Burn and Elvanfoot. Also a road through the County of Lanark, the purpose of which was to connect Falkirk and the other great cattle markets of Crief and Doune with Carlisle. In addition, Telford made no less than three surveys for an improved road from Carlisle to Port Patrick, but although he repeatedly urged its importance as a route to Ireland (Port Patrick–Donaghadee, twenty-two miles, is the shortest sea crossing) the work was not carried out in his lifetime.

It was on the line of this road, however, that Telford's first Scottish bridge was built across the Dee at Tongueland in Kirkcudbright (1805–6). This is an impressive stone bridge having a single arch of 112 feet span. It was the first of Telford's road bridges to be built with hollow spandrels. It is also unusual in possessing no rise to the crown of the arch. On almost all Telford's road bridges the roadway rises to the crown of the centre arch, his chief reason for this being that rainwater could not lie on the bridge but would quickly run off. In addition to Tongueland there are a number of imposing bridges on the Glasgow and Falkirk roads, but by far the most dramatic in the Lowlands, if not in all Scotland, is the Cartland Crags bridge near Lanark. Its three graceful stone arches stride across the rocky gorge of the Mouse Water at a height of 129 feet above the stream. It was completed in 1822.

It would be a mistake to suppose that when Telford had completed his plans, surveys and bridge designs he then delegated all responsibility to Mitchell or to Gibb and their subordinates. On the contrary, notwithstanding the works that constantly called for his attention in England, not a year went by without his making at least one extended tour of the Highland works. On his tour of 1812 he was accompanied by a youthful engineer, William Provis, whom he was training and who was soon to become one of his most able assistants. In 1819 he had for companion none other than Robert Southey, the poet laureate, whose acquaintance he made through their mutual friend John Rickman.

Southey posted up from his home, Greta Hall, Keswick and joined forces with Telford at Edinburgh in the middle of August. Although seventeen years separated them in age, the two men took an immediate liking to each other. 'There is so much intelligence in his countenance,' wrote Southey, 'so much frankness, kindness and hilarity about him, flowing from the never-failing well-spring of a happy nature, that I was upon cordial terms with him in five minutes.' How genuine that liking was was proved during the next six weeks travelling through the Highlands when they were never out of each other's company, often sharing great discomforts and the same bedroom in many indifferent and primitive Highland inns. Such an experience will either break a friendship or cement it for ever; there can be no half-measures.

As a poet, Southey cannot stand comparison with his brother-in-law Coleridge or his other great contemporaries. This has tended to obscure the fact that when he put off the mantle of the Laureate, forgot to be pompous, and wrote unselfconsciously, he became a different person, revealing a highly intelligent, lively mind with acute powers of observation and shrewd judgement. This is revealed, not only in his incomparable letters but also in the diary which he kept throughout his tour with Telford. Reading it we find, besides revealing glimpses of Telford and his associates, a picture of the social revolution which was taking place so swiftly in the Highlands as a result of the new roads, new bridges and new harbours, many of which had been completed by this date. When, on 1 October at Longtown near the Scottish Border the two men parted and went their separate ways, Telford to Edinburgh and Southey back to Keswick, Southey wrote in his diary: 'This parting company, after the thorough intimacy which a long journey produces between fellow travellers who like each other, is a melancholy thing. A man more heartily to be liked, more worthy to be esteemed and admired I have never fallen in with; and therefore it is painful to think how little likely it is that I shall ever see much of him again ...'

6

THE CALEDONIAN CANAL

THE proposal for a canal through the Great Glen of Scotland was no new one when Telford advanced it in 1801. This is not surprising for, as a glance at a map shows clearly, this great natural rift cutting diagonally through the Highlands from sea to sea seems deliberately to invite the attention of ambitious canal engineers. For although it is 113 miles long, 52 miles of this distance consist of sea loch, while the fresh-water lochs, Ness, Oich and Lochy account for a further 38 miles, leaving only 23 miles of canal to be cut. Such a passage had been mooted for many years and the first man to carry out a survey was James Watt. Watt did this work for the Commissioners of Forfeited Estates in 1773 when he was practising as a land surveyor in Glasgow before his steam engine brought him fame. Watt's estimate for a canal 10 feet deep with 32 locks 90 feet long and 25 feet wide was £164,000. In 1793, John Rennie consulted Watt and prepared a second scheme, but again nothing was done.

Telford in his turn consulted Watt about his survey before he made his first Highland Report in 1801. The two men were already acquainted, for Telford had had dealings with the firm of Boulton & Watt in 1795 over the supply of a thirty-inch beam engine which was installed near Chester to pump water into the Wirral line of the Ellesmere Canal. Although Telford now proposed a canal of greater dimensions, his survey closely followed the line proposed by Watt. This does not mean to say, however, that he blindly accepted Watt's conclusions. On the contrary he surveyed the project with great thoroughness before he submitted his recommendations. As a good canal engineer, one of his first concerns was to assure himself that there would be an adequate supply of water for the summit level. The summit in this case would be Loch Oich. This is a small and shallow loch, but the river Garry flows into it at Invergarry and Telford rode up Glen Garry in order to satisfy himself on this vital point. He came to the conclusion

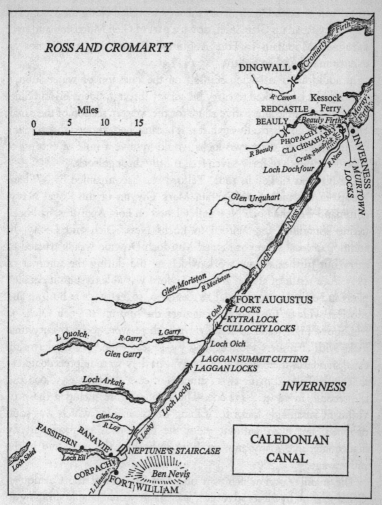

ROSS AND CROMARTY

DINGWALL

R. Conon · Kessock
REDCASTLE · Ferry
BEAULY · Beauly Firth
PHOPACHY
R Beauly · CLACHNAHARRY
Craig Phadrick
Loch Dochfour · R Ness

INVERNESS
MUIRTOWN
LOCKS

Miles
0   5   10   15

Glen Urquhart

Loch Ness

Glen Moriston · R Moriston

Glen Moriston · FORT AUGUSTUS
R Oich · LOCKS
KYTRA LOCK
CULLOCHY LOCKS

L. Quoich · R. Garry · L. Garry
Glen Garry · Loch Oich

LAGGAN SUMMIT CUTTING
LAGGAN LOCKS

Loch Arkaig

INVERNESS

Loch Lochy

Glen·Loy
R Loy · R Lochy
FASSIFERN · BANAVIE
Loch Shiel · Loch Eil · NEPTUNE'S STAIRCASE

CORPACH · Ben Nevis
L. Linnhe · FORT WILLIAM

CALEDONIAN
CANAL

that this river, with the two lochs, Quoich and Garry, on its course
acting as natural reservoirs, would be a perfectly adequate feeder.
Having reached the head of Glen Garry, Telford's own account of his
subsequent movements is worth quoting because it gives us a glimpse of
the kind of travelling his Highland surveys involved. 'I passed', he
writes, 'by a very rocky and precipitous track down to the head of
Loch Hourn; from Loch Hourn I travelled by a track scarcely less
rugged to the top of Glen Elg and over the steep mountain of Raatachan

to the top of Loch Duich; from thence I travelled along the vestiges of a Military Road up Glen Shiel, down a part of Glen Morriston and over a rugged Mountain to Fort Augustus.' Even today this is not an excursion to be taken lightly.

In addition to satisfying himself on the question of water supply, Telford had been asked to carry his survey through Lochs Eil and Shiel as being a possible alternative route for the western section of the canal. He reported emphatically against this because he reckoned that to cut a channel between these two lochs would involve a mile of cutting of an average depth of forty-seven feet, mostly through rock.

On his next survey in 1802, Telford was accompanied by William Jessop. In the meantime Captain Mark Gwynn of the Royal Navy, commander of the Loch Ness galley based on Fort Augustus, had been taking soundings for Telford in Lochs Ness, Oich and Lochy. In addition, a local surveyor named Murdoch Downie was instructed to carry out further survey work which he did during the summer of 1803. As a result of all this activity Telford was able to submit detailed plans and estimates for a canal 20 feet deep, 50 feet wide at bottom and 100 feet wide at top. To carry it over the summit at Loch Oich, 94 feet above high water in Beauly Firth, there were to be twenty-nine locks with chambers 170 feet long by 40 feet wide.[1] By modern ship canal standards these figures are puny, but they were unprecedented at that time. Telford made two estimates of cost, the first £350,000 and the second, in 1804, £474,000. The second figure included the provision of small side locks for fishing boats, an idea which was soon dropped, but even without them the figure was to prove utterly inadequate, like many another estimate both in the canal and in the railway age.

The main reason which was urged in support of the Caledonian Canal scheme by all its advocates was that it would enable shipping to avoid the dangerous and difficult passage round the northern coast of Scotland through the Pentland Firth. This route was used by ships trading from east coast ports to North America and by those engaged in the important Baltic timber trade between ports on the western coast of England and Memel. The example was quoted of two ships which left Newcastle on the same day, one bound for Bombay via the

1. As built the locks can pass craft measuring 150 ft. × 35 ft. on a draught of 13 ft. 6 in. or 160 ft. × 35 ft. on a draught of 9 ft. 6 in.

English Channel and the Cape of Good Hope and the other for Liverpool via the north of Scotland. The latter was so long stormbound that she spent longer on the voyage to Liverpool than the other took to reach Bombay. In winter, ships sometimes lay stormbound at Stromness in Orkney for as long as two months. Added to this was the secondary reason that the building of such a canal would be a great new source of employment in the Highlands and would therefore check emigrations. These arguments had lost none of their force by 1801, but what really caused Telford to succeed where Watt and Rennie had failed was the fact that Britain was at war with France. It was felt that the canal would not only protect British merchant shipping from the attentions of French privateers, but would also be a valuable strategic route for ships of war. This argument proved decisive although those who advanced it must either have had a pessimistic view of the duration of the war or else a very optimistic notion of the time it would take to build the canal.

If Telford himself encouraged such hopes of speedy completion he should have known better, for he was proposing to carry out an engineering work of prodigious magnitude under conditions which could not have been more difficult. The country was wild and inhospitable in the extreme, the weather often violent and the local inhabitants quite unused to organised labour on such a scale. Added to this, communications were non-existent and practically all the equipment and materials required would have to be brought by sea over great distances. However, the die was cast, the Caledonian Canal Commission was set up and Telford was ordered to proceed with the work.

He decided to concentrate first on the building of the sea locks and harbour basins at the eastern and western terminals of the canal; at Clachnaharry on Beauly Firth near Inverness and at Corpach at the head of Loch Linnhe near Fort William in the west. He appointed two resident engineers for these eastern and western districts, Matthew Davidson (Clachnaharry to Loch Ness, 7¾ miles) and John Telford (Corpach to Loch Lochy, 8 miles). John Telford was apparently no relation of Telford's, though he must have been a kinsman. He had been appointed toll collector on the Ellesmere Canal Company's Wirral Line at Chester in June 1797, where he must have subsequently distinguished himself to merit such an important promotion. Unfortunately he died in 1807; his widow returned to Chester at the

Commission's expense and he was succeeded by Alexander Easton. In practical charge of the works as contractor was John Simpson with John Wilson and John Cargill working as foremen masons under him in the western and eastern districts respectively.

Telford thus assembled the same splendid team which had served him so well in building the aqueduct at Chirk and the great piers of Pont Cysyllte, but here in the Highlands they had the greatest difficulty in training and welding together an effective labour force, especially at the more remote western end of the canal. Houses for the masons and a barrack building for labourers had first to be built at Corpach. Four hundred bolls of oatmeal were put in store at Fort William to feed the workmen and a brewery was built at Corpach so that, in the words of the Commissioners, 'the workmen may be induced to relinquish the pernicious habit of drinking Whisky'. Notwithstanding these preparations, labour troubles at the outset almost drove poor John Telford to distraction. 'Last Saturday', he told Telford soon after work had begun in 1804, 'was pay day and a very disagreeable one it was; notwithstanding the men was all informed when you was here that those upon days wages would only receive 1/6d pr day, they refused to take it. Nor do I suggest it will be settled without going before a Justice. Mr. Wilson and myself were in eminent [*sic*] danger of our lives; yet notwithstanding we would not give way to one of them, tho they threatened much and were on the point of using violence several times...'

Things seem to have been little better a year later, for on an occasion when the wages money had failed to arrive at Corpach, John Telford sent a desperate appeal for funds to Davidson at Clachnaharry. 'If the men are not all settled with Monday night at farthest,' he wrote, 'I dread the consequences.' And he adds in a postscript: 'We are all well here at present – God knows how long we shall remain so if John does not come here with the money on Monday night.' His chief was more philosophical, perhaps because he was not in the direct line of fire. 'Misunderstanding and interruptions', Telford warned Rickman, 'must be expected in works of this kind amongst a people just emerging from barbarism.'

Matthew Davidson was an older man than John Telford and a much tougher character. He was indeed a very remarkable character in every way, as is revealed by the letters which he wrote to his sons. Southey described him as a strange, cynical humorist and said that in person and manners he much resembled Doctor Johnson. Like Telford he was a voracious reader, so much so that he became known as the Walking Library, yet towards the end of his life he remarked to his son that the Bible was the only book that deserved reading, 'all the rest is only looking at objects through a blind man's spectacles'. Davidson never toyed with republicanism as did Telford but remained throughout his life a staunch advocate of the established order. Extremists, demagogues or absolutism in any form could always be relied upon to provoke a retort such as this (in 1809): 'England these 120 years has been the best governed nation ever existed on this Globe and yet the factious barbarians are never content. Hang the ringleaders and banish the rebel mob to the Highlands of Scotland – this would tame them if anything would.' And when one of his sons suggested that he seemed to disapprove of any opposition whatever to established power he replied: 'I am equally hostile to unbridled power whether exercised by the head or the tail of Society. The last is the most hideous and most to be dreaded in these times.' One of his more picturesque eccentricities was his belief in what he called 'that elegant and classical medicine, Bathing in cold water' as a cure for every mortal ill. He would astound all beholders by bathing in the Beauly Firth in the most inclement weather, and when his son John was dangerously ill with typhus he was only narrowly dissuaded from plunging the unfortunate boy into a bath filled with cold sea water. On another occasion he threw two bucketfuls of his infallible specific over the Davidsons' serving maid when she complained of a fever. He claimed an instantaneous cure. It is a comment on the changing times that in 1812 he suddenly announced to his son Tom that he proposed to destroy his wig: 'Scratch wigs are going out of fashion very fast – the first moderate day mine goes ... grey hairs are more suitable than sham auburn at fifty-eight.'

Being a Lowland Scot, Davidson affected a fine contempt for the Highlands and its people. John Telford maintained that he would not accept a seat in heaven if there was a Highland man in it, while Southey quoted him as saying that if justice were done to the inhabitants of Inverness, in twenty years there would be no one left there but the Provost and the hangman. When he came south to join Telford he had soon learned to love the people of Wales and married a Welsh wife who bore him his three sons before he left for the Highlands. Two,

Thomas and John, trained as doctors, but the youngest, James, was his darling and succeeded him at Clachnaharry after his death. 'James', he wrote, 'has all the warmth of Cambria in his constitution; I love him for his feeling heart.' He often remembered with regret his old home from whose windows he had watched the great stone piers of Pont Cysyllte slowly rising in the valley below. Sent a new telescope by Thomas he replied: 'I lament much the want of the Vron or Trevor Hill for a fine view into England.' He brought with him to Clachnaharry so many of the Welsh masons who had worked under him at Pont Cysyllte that his house became the centre of a colony of exiles; indeed he always referred to them as 'the colony' as though they were some outpost in darkest Africa. 'There is such a noise of Welshmen in the kitchen – the battle of Borodino was nothing to it', he once wrote. Messages would be sent south from these exiles for delivery by his son Thomas who was studying with a doctor in Oswestry, thus: 'If you have an opportunity, let Thos Jones, Mason, at Llanfair Caerinion, Montgomeryshire know that his son Thos and family are well. This country is not so pleasant as his own but he has better wages, works as he likes and we use him as kindly as circumstances will admit.'

Matthew Davidson and his Welshmen were set a task at Clachnaharry which was less spectacular than the great aqueducts they had built on the Welsh border but far more difficult. The shore of the Beauly Firth at Clachnaharry shelves very gradually. This meant that the entrance lock for the new canal must be carried out 400 yards beyond the shore line in order that ships could enter it at any state of the tide. On the site chosen for this lock an iron bar was driven into the mud; it sank fifty-five feet before it found solid bottom. The consistency of this mud was such that oak piles rebounded from it as though they were on springs at each blow of the driver and all hope of constructing a coffer-dam had to be abandoned. The solution adopted was to build out from the shore a great clay embankment. On the site of the lock this was weighted with stone and left for six months during which time it slowly settled, extruding the soft mud from beneath it. When this process of consolidation was complete, piles were driven and great oblong timber frames secured to them. These formed the main members of a huge coffer-dam within which the artificial earthwork was excavated to the full depth of the lock chamber. Horse-driven chain pumps and a 9 hp. Boulton & Watt beam engine worked constantly to keep the lock pit clear of water. In the bottom of the excavation a bed of rubble stone bound in water-lime mortar two feet thick at the centre and five feet thick at the sides was laid down as a foundation for the lock chamber invert and the side walls which were six feet thick. It was a work of the greatest difficulty which was not completed until 1812, but it proved perfectly satisfactory. To build successfully the largest lock in the world at that time in the middle of a morass of mud was indeed a triumph for Telford, Davidson and his Welsh masons.

In forming the great embankment and for conveying stone to this and to the other locks at Clachnaharry and Muirtown a considerable length of iron railway was laid down. This was supplied by Leys Ironworks at Aberdeen and from Jessop's Butterley Ironworks in Derbyshire. Ironwork for the locks in the eastern district also came from Butterley, being shipped on to the Trent at Gainsborough and thence by sea from Hull. With the exception of the Beauly sea lock where North Wales oak was used, the lock gates themselves were of cast iron with sheathing of Memel pine. The sloop *Caledonia* was built to the Canal Commissioners' order by Samuel Deadman of Inverness and launched in March 1804 for the purpose of ferrying materials from the east coast ports to Clachnaharry.

In finding stone for building his locks, Matthew Davidson was lucky. There were suitable beds of rubble stone in the slopes of Craig Phadrick which overlook the Firth at Clachnaharry, and 1,100 yards of railways was sufficient to run this stone direct from quarry to lock sites. Freestone for facings and quoins was quarried at Redcastle on the Black Isle and ferried over the Firth. Limestone suitable for mortar making was found near Phopachy, three miles along the shore of the Firth.

The making of the second lock at Clachnaharry and of the basin and four locks at Muirtown did not occasion any especial difficulty, though they were works of such size that they took many years to complete. After the Beauly sea lock, Davidson's most difficult undertaking was the regulating lock[1] situated where the eastern arm of the canal enters the little Loch Dochfour which forms the eastern extremity of Loch

1. In any canal which unites a series of lakes, regulating locks are essential at each junction between canal and lake. Their function is to prevent the level of water in the canal from being affected by any variations in the lake levels due to drought or flood.

Ness. Here the river Ness had to be diverted from its course for a distance of half a mile and the lock chamber was then built in the old bed of the river. As this chamber had to be sunk to a depth of twenty feet below the level of the river this was a delicate and hazardous operation.

Davidson just lived long enough to see his portion of the work completed. In June 1818, a few weeks before his death and already a very sick man, he was able to write to his son Thomas, now a fully fledged surgeon at Nottingham: 'Vessels are now daily and regularly passing to and from Fort Augustus with Stone, Coal, meat and provisions &c., there were one day last week 10 Vessels on Loch Ness at the same time. They can make the passage out and in in 3 days, and two of them made the Voyage in two days. This is the way we arrange matters here. Do you wish a trip to Fort Augustus?'

In building the great western sea lock at Corpach, John Telford and his successor Alexander Easton wrestled with difficulties of quite a different kind. While Davidson and his men wallowed in an almost bottomless quagmire, they had to cut the excavation for their chamber out of solid rock. They had more trouble with water than did Davidson, and the cavernous chamber far below the level of Loch Linnhe where the temperamental Highland labourers hacked and blasted their way through the rock was only kept clear by the ceaseless efforts of a 20 h.p. Boulton & Watt beam engine. This had been sent up in parts from the famous Soho Works at Birmingham by canal and sea and Taylor, one of Boulton & Watt's fitters, was dispatched to Corpach to supervise its erection. This must have been no mean feat in that remote place. The bigger parts of this engine were shipped at Chester by the new sloop *Corpach* which had been built there by William Cortney & Co. to the Commissioners' order for duty in the western seas. But certain boxes of small parts, presumably the valve gear, were sent later 'by Gilbert & Worthington's boat' to Preston Brook for shipment via Runcorn and the Mersey, for we find Telford writing to Boulton & Watt in June 1805 complaining bitterly that the boat, with its precious boxes, had been sunk at Runcorn. Such were the transport problems and troubles of those days.

A second sloop for the western distict (also named *Caledonia*) was built at Fort William by John Stevenson because the demand for sea transport at Corpach was much greater than at Clachnaharry. As at the eastern terminal, a railway was laid from a rubble stone quarry at Banavie, but all other materials had to come by water. More rubble stone from Fassifern on Loch Eil; granite from a quarry on Loch Linnhe near Ballachulish; limestone, to be burnt in a kiln at Corpach, from Sheep Island in the Firth of Lorne. No nearer source of freestone suitable for lock quoins was found than the island of Arran in the Cumbraes. Iron for railways, locks and swing bridges on the western section came from Hazledine's foundry at Plas Kynaston and from Wilkinson's Bersham ironworks. As at Clachnahrry, the Corpach sea lock gates were of Welsh oak the better to resist the action of salt water.

Two locks lifted the canal out of the basin at Corpach and then, after a mile of level cutting, came the great flight of eight locks at Banavie which became known as Neptune's Staircase. Midway between Banavie and the western end of Loch Lochy an aqueduct of three arches had to be built to carry the canal over the river Loy and farm accommodation roads on each bank of that river. Here, too, a saw mill was built driven by the waters of the Loy. Great care had to be taken on this section to control by sluices and culverts of generous size the waters which swept down from the mountains in seasons of heavy rain or melting snows. At one point three discharge sluices were built through which flood-water could be sent thundering down into the deep bed of the Lochy far below out of an arch twenty-five feet high built into the rock. 'What would the Bourbons have given for such a cascade at Aranjuez or Versailles!' Southey exclaimed when he saw these sluices in action in 1819. 'The rush and the spray and the force of the water reminded me more of the Reichenbach than of any other fall.' He was even more impressed by the great flight of locks at Banavie, calling it 'the greatest piece of such masonry in the world and the greatest work of its kind, beyond all comparison'.[1]

Early in 1818, as the eastern division neared completion, John Cargill moved his headquarters from Clachnaharry to Fort Augustus, where he built a house for himself which was described as the only decent house in the place. By the time Matthew Davidson died practically all his Welsh colony had followed Cargill thither. As the western division

1. Southey evidently had these locks in mind when he wrote the lines quoted at the beginning of this book, for boats are not 'upraised eight times' at Clach-naharry as his lines imply.

also approached completion the same process took place, so that by 1819/20 almost the whole labour force from east and west was concentrated upon the remaining eleven miles of the central section of the canal between Fort Augustus and the east end of Loch Lochy. The eastern force under Cargill took the length from Fort Augustus to Loch Oich, while John Wilson's men were responsible for the portion from Loch Oich to Loch Lochy which included the great summit cutting at Laggan.

It was necessary to deepen the navigable channel through Loch Oich, the approach to the bottom of the flight of five locks at Fort Augustus and the bed of the river Oich above the locks from which, as at Dochfour, the course of that river had been diverted. For this work two of the first steam bucket dredgers ever built were used. They were constructed at the Butterley Ironworks to the designs of that very eminent and versatile engineer, Bryan Donkin. The first arrived in parts in 1816 and was assembled on the shores of Loch Oich by Thomas Rhodes, an engineer who was responsible for erecting much of the ironwork on the canal and whom we shall meet again in another chapter. The second was landed at Fort Augustus in 1818, the canal having then been opened thus far. These two machines cost over £6,000.

The greatest difficulty was encountered in building the bottom lock chamber at Fort Augustus. This had to be sunk far below the level of Loch Ness and a bed of loose, permeable gravel was encountered through which the waters of the loch poured in almost uncontrollable quantity. Here the last and largest of the three Boulton & Watt engines which had been supplied to the Commissioners was pressed into service. This was of 36 h.p., having a cylinder 4 feet in diameter and a piston stroke of 8 feet – hardly a handy unit to haul about the Highlands. Yet even this titan, labouring night and day, could not cope with the constant influx and finally both the Clachnaharry and Corpach engines as well had to be pressed into service before the waters were mastered and the chamber built.

By their very remoteness, Telford's great works in the Highlands never excited the public wonder and acclaim which works of far smaller magnitude aroused in populous England. Thus the mighty summit cutting at Laggan by which Telford drove the Caledonian Canal from Loch Oich to Loch Lochy dwarfs in its scale almost all the works of the later railway builders which the world was soon to marvel at. Railways and barrow runs were both used in the work and as soon as a part of the excavation was deep enough to be flooded the Donkin dredging machines were brought in to increase the depth. But few visitors saw the Laggan cutting while it was being driven, and when it was finished the water which filled it and the trees which were planted to bind its sides soon concealed the scale of it. Fortunately, however, we have a description of the work, both here and at Fort Augustus, as seen through the eye of an intelligent layman, in Robert Southey's diary of his tour with Telford in 1819. He and Telford arrived at the inn at Fort Augustus on the evening of Thursday 16 September, and on the following day he wrote: 'Went before breakfast to look at the Locks, five together, of which three are finished, the fourth about half-built, the fifth not quite excavated. Such an extent of masonry, upon such a scale, I have never before beheld, each of these Locks being 180 feet in length. It was a most impressive and rememberable scene. Men, horses and machines at work; digging, walling and puddling going on, men wheeling barrows, horses drawing stones along the railways. The great steam engine was at rest, having done its work. It threw out 160 hogsheads per minute; and two smaller engines (large ones they would have been anywhere else) were also needed while the excavation of the lower locks was going on; for they dug 24ft below the surface of water ... and the water filtered thro' open gravel. The dredging machine was in action, revolving round and round, and bringing up at every turn matter which had never before been brought to the air and light. Its chimney poured forth volumes of black smoke, which there was no annoyance in beholding, because there was room enough for it in this wide clear atmosphere. The iron for a pair of lockgates was lying on the ground, having just arrived from Derbyshire: the same vessel in which it was shipt at Gainsborough, landed it here at Fort Augustus. To one like myself not practically conversant with machinery, it seemed curious to hear Mr. Telford talk of the propriety of weighing these enormous pieces (several of which were four tons weight) and to hear Cargill reply that it was easily done.

'After breakfast we went to inspect the works in progress between this place and Loch Lochy. This was a singularly curious and interesting sight. What indeed could be more interesting than to see the greatest work of its kind that has ever been undertaken in ancient or modern times, in all stages of its progress – directed everywhere by perfect skill and with no want of means. . . .'

Telford and Southey continued on their way westwards until they encountered the second dredger working on Loch Oich and the poet writes: 'At this (the Eastern) end of Loch Oich a dredging machine is employed, and brings up 800 tons a day. Mr. Hughes [one of the Clachnaharry Welsh colony], who contracts for the digging and deepening, has made great improvements in this machine. We went on board and saw the works; but I did not remain long below in a place where the temperature was higher than that of a hothouse, and where machinery was moving up and down with tremendous force, some of it in boiling water.'

Finally they came to the Laggan cutting and Southey writes: 'Here the excavations are what they call "at deep cutting", this being the highest ground in the line, the Oich flowing to the East, the Lochy to the Western sea. . . . The earth is removed by horses walking along the bench of the Canal, and drawing the laden cartlets up one inclined plane, while the emptied ones, which are connected with them by a chain passing over pullies, are let down another. This was going on in numberless places, and such a mass of earth had been thrown up on both sides along the whole line, that the men appeared in the proportion of emmets to an ant-hill amid their own work. The hour of rest for men and horses is announced by blowing a horn; and so well have the horses learnt to measure time by their own exertions and sense of fatigue, that if the signal be delayed five minutes, they stop of their own accord without it. . . .'

As Southey's account indicates, the works were at last within sight of completion. The previous year Telford had said that he hoped to open the canal for through passage at the latter end of 1820 or early in 1821, but the difficulty at Fort Augustus defeated this forecast and it was not until 23 and 24 October 1822 that the first boat was able to sail through the Great Glen from the eastern to the western sea. So far as Telford was concerned it was a hollow victory. The canal had cost over £900,000 and instead of the seven years he had estimated, eighteen had been spent in building it. Moreover, it was still far from perfect for instead of the twenty-feet draught originally specified, the depth of channel through Loch Oich and through the summit cutting was only twelve feet despite all the efforts of Donkin's dredgers.

To make matters worse, during these years circumstances had changed and developments had taken place all of which militated against any hope of commercial success for the canal. The Napoleonic War had, of course, become a memory so that the canal no longer possessed any strategic value. The size of shipping was increasing and many new naval and commercial craft had already outgrown Telford's locks even if the depth of the canal had been adequate. The steamship, too, was coming over the horizon. One of Henry Bell's paddle steamers began plying on the canal between Muirtown and Fort Augustus as early as 1820, while some years later the introduction of four steam tugs eased one of the canal's difficulties – that of sailing vessels being delayed, particularly on the east to west passage, by headwinds in the lochs. In these ways steam power helped the canal but in the long term it was fatal because it robbed of its old terrors the northern passage through the Pentland Firth. But the unkindest cut of all was the crippling duty imposed by the Government, with the object of favouring Canadian timber, on all deals imported from the Baltic. This robbed the canal of its chief source of commercial revenue. Telford wrote bitterly that the loss on the canal against which the Government inveighed was 'a trifling national detriment indeed compared to the vast damage inflicted by the compulsory substitution of bad timber in place of good timber for all the many purposes to which deal-balk is convertible. The separation of Canada from the mother country may perhaps hereafter remedy this evil.'

Much of the criticism of the canal by public and parliament was directed against Telford, though he seems to have been far less perturbed by it than was John Rickman. It was scarcely fair to Telford since he could not have been expected to foresee the future, nor was it his job to try to do so. He had, it was true, been guilty of a gross underestimate, but the swollen cost of the canal was in a great measure due to the vast increase in wages and prices which rocketed as a consequence of the Napoleonic Wars. Commercial considerations were not Telford's concern; his job had been to build the canal and much more serious from his point of view were the engineering defects which soon developed. There were slips in Laggan cutting and troubles with some of the lock walls which culminated when the entire side wall of the bottom lock at Fort Augustus, which had caused so much trouble in building, blew in. The fact that most of these defects were in the central division suggest that work may have been scamped there in the effort to complete the canal in response to the constant pressure from above. It may also have been due to the absence of one very keen pair of eyes – those of John Simpson, Telford's 'treasure of talents', who returned home to die at Shrewsbury in June 1816. Telford must have felt the loss of Simpson and Davidson very keenly, and the canal work must have suffered from their absence. Such men are not easily replaced.

By the time these defects had been put right over a million pounds had been spent on the canal, and it was still so far from satisfactory that by 1840 it was in a state of dilapidation. After attempts to sell it to private enterprise had failed the question of restoration or abandonment was hotly debated in parliament. The restorers won the day and the canal was re-opened to traffic in May 1847, its depth having been increased to seventeen feet.

Despite its melancholy history the Caledonian Canal stands open to this day as a monument of extraordinary engineering achievement and tenacity of purpose having regard to the period and the conditions in which it was built. Moreover, on one occasion at least, Telford's great canal has served this country well. This was in 1918 when it was decided to lay a great mine barrage across 235 miles of the North Sea from the Orkneys to Norway in order to defeat the German U-boat menace. Seventy thousand mines, supplied in parts from America, were all shipped through the Caledonian Canal to Clachnaharry, where they were assembled at depots at Dalmore and Muirtown. This involved 6,254 passages through the canal in twelve months, and gives some idea of the traffic the canal could pass when needed. So a water-way which never fulfilled its intended purpose in the war with Napoleon lived on to justify itself and its creator in that later and greater war against Kaiser Wilhelm.

7

THE GOTHA CANAL

TELFORD's climb to fame ended in 1805, twenty-three years after he left Eskdale to seek his fortune. The completion of the Pont Cysyllte aqueduct in that year, coinciding as it did with the launching of his great projects in the Highlands, set him in a position of such eminence in his profession that he possessed only one rival of equal stature – John Rennie. He had reached, not a peak but a level plateau of fame, for at forty-eight his physical and mental powers were still those of a young man and there lay before him many years of great endeavour.

It must have been very gratifying to him to discover that the fame of his engineering exploits was not only nation-wide but had spread to the continent of Europe. That this was so was brought home to him by a letter which reached him when he was on a visit to Clachna-harry at the end of May 1808. Like most of Telford's correspondence it had been chasing him round the country. Originally addressed to him in Edinburgh it had been forwarded successively to Ellesmere and Shrewsbury before it finally caught up with him at Inverness. It proved to be a formal communication from King Gustav Adolf of Sweden requesting his assistance in the survey and construction of a ship canal from the North Sea to the Baltic. Enclosed with this impressive document was an informal and much more informative letter from a certain Count Von Platen. This began: 'You will, I hope, excuse me for introducing myself to you in this manner and I fear in English not of the best,' and went on to explain the nature of the project in some detail and to beg for Telford's co-operation. 'You may judge then, sir,' the Count concluded, 'how anxiously I look for you after having uninter-ruptedly for more than four years worked upon to bring it so far as it is now.' Few engineers could have resisted this combination of grandiose Royal edict with quaintly worded personal appeal, and Telford replied to Von Platen on 2 June signifying his willingness to assist.

The idea of building a waterway across Sweden was as old as that of the Caledonian Canal. Although there was here no natural defile comparable to the Great Glen and the distance from sea to sea was very much greater, the existence of the two great lakes Vänern and Vättern as well as a number of smaller lakes prompted the idea of stringing them together on the thread of a man-made canal. As in the case of the

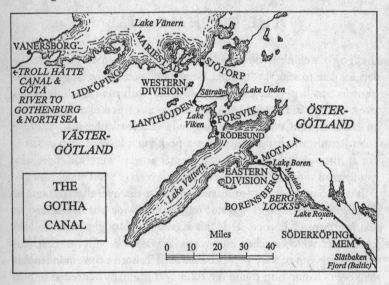

Caledonian, the arguments in favour of the Swedish project were partly commercial and partly strategic. All shipping bound to or from Stockholm and other ports on Sweden's Baltic seaboard must necessarily pass from the Kattegat to the Baltic by the extremely narrow sound which separates Sweden from Denmark and which is dominated by the ancient castle of Elsinore. An alternative route to the North Sea from Stockholm was not only desirable for the Swedish navy; it would also avoid payment of the heavy tolls levied by Denmark on all commercial shipping passing through the Sound. Finally, by linking her great lakes with the eastern and western seas a prodigious area of Sweden would be brought within range of water transport.

After several false starts and abortive efforts, the first and westernmost link of this coast to coast waterway was completed in 1800. This was the Trollhätte Canal which links Lake Vänern with the navigable river Göta which falls into the North Sea at Gothenburg. An engineer named Samuel Bagge was in charge of this work and one of the directors of the Canal Company was Admiral Count Baltzar Bogislaus von Platen, Knight of the Sword and a Member of the Swedish House of Nobles.

Fired by the immediate success of the Trollhätte Canal, von Platen determined to revive the scheme for the complete line of canal through to the Baltic which had last been surveyed between 1781-4 by two Swedes, Danial av Thunberg and Elias Schweder. That the Count would succeed in this where others had failed was due to his political influence, his genius for organisation, and, above all, to his passionate determination. Once he had resolved to drive the Gotha Canal through to the Baltic, its realisation became the sole aim and purpose of his life, a cause to which he dedicated himself utterly. He had heard of the great Caledonian Canal project and he envisaged the two schemes as being complementary to one another. Together, the two canals would form parts of a new shipping route between the Baltic and the western ports of England. What better, then, he had argued, than that the same engineer should be associated with both these great canals? It was therefore as a result of four years tireless lobbying on von Platen's part that the Gotha Canal project received the blessing of King and Government and that the invitation was sent to Telford.

The suggested visit was complicated by the international situation and in his letter of acceptance Telford pointed out that: 'these Northern seas are at present much infested with Danish and French privateers. This makes a passage in a common trading Vessel to be attended with a risk from these armed Vessels however trifling. I must therefore stipulate that I shall be taken up at Aberdeen and also landed at the same place by either a Swedish or English ship of War. . . .' The indefatigable Count soon solved this difficulty by paying a personal visit to the English Admiral Sir James Saumarez who was then aboard his flagship *Victory* in Gothenburg habour. Here he 'meeted the most polite reception', as he put it, and was able to arrange for Telford to travel by armed convoy from Leith.

It was from Leith, then, that Telford set sail for Sweden on 28 July 1808 accompanied by two assistants, William Hughes and Henry Fulton. Strangely enough it was Telford's first sea voyage and he evidently decided to take no chances, laying in for himself and his two companions enough stores to last them for an Atlantic voyage. They included three hams, each weighing 48½ lb., ¾ cwt. of biscuits and

31¼ lb. of lump sugar. The liquor supplies were even more lavish and show that with the coming of success Telford had abandoned the teetotalism and the 'sowens and milk' simplicity of his early Shrewsbury days. They included two dozen each of Madeira and port, half a dozen each of gin and brandy and six dozen of porter. In the event, the passage was uneventful and so speedy that one is left wondering just how much of this prodigious quantity of liquor the intrepid trio had managed to put away by the time they landed at Gothenburg only six days after leaving Leith.

Telford met von Platen for the first time on 8 August and the two men took an immediate liking to each other, recognising instantly, no doubt, that despite differences of nationality, birth and upbringing they were birds of very similar feather. The many letters which the Count wrote to Telford over the years to come declare in the warmest terms the deep admiration and affection which he felt for him. As we should expect, Telford's replies are shorter and written with more reserve, but that the regard was mutual is revealed by the frankness with which he takes the Count into his confidence by detailing the engineering projects which he has in hand.

This first meeting was followed by six weeks of most intensive and tireless survey work in the course of which Telford covered the whole line of the proposed canal, a distance of 114 miles, and fixed the sites for the locks. He then spent a week at von Platen's manor of Frugården drawing up his plans and report before returning to Gothenburg. He was still working on his report until the moment came for him to board the *Diana* bound for Harwich on 1 October. The basis of Telford's survey was Thunberg's original line, although he advised deviations in many places and recommended more locks of greater length and breadth but lesser individual fall. Doubtless because of the greater magnitude of the work he proposed a waterway upon a scale smaller than that of the Caledonian, the dimensions given being depth 10 feet, width at bottom 42 feet and at top 82 feet, the lock chambers to be built to pass vessels of a maximum size of 105 feet in length by 23 feet beam. There would be 53 miles of new canal to cut. This, with 133 miles of lake navigation and the existing 52 miles of navigation between Gothenburg and Lake Vänern, made up a total of 238 miles from sea to sea.

The great Lake Vättern divided the proposed canal into two distinctly

separate western and eastern sections which were situated in the provinces of Västergötland and Östergötland respectively. The western
line was to commence at Sjötorp on the eastern shore of Lake Vänern
and run to Rödesund on the west shore of Lake Vättern. Although it
was the shorter of the two sections it was to prove a work of great
difficulty. It included the summit level of the whole system, 278 feet
above sea-level, at Landhöjden immediately west of the narrow,
winding Lake Viken through which the navigation was to pass. Here,
in wild and remote country, a very considerable cutting through
granite rock had to be driven and this proved to be a work of such
difficulty that in places the width of channel had to be reduced to
twenty-four feet. Fortunately, just as Telford had found in the river
Garry and its lochs an adequate feeder for the summit of the Caledonian Canal at Loch Oich, so here he was satisfied that the river
Sätraån, which flows out of Lake Unden, would provide an ample
natural water supply to this much higher top level. It was at the hamlet
of Sätra by the bank of this river that Telford, von Platen and their
assistants made their headquarters while they were surveying this
western line.

The eastern line of the Gotha Canal was projected to run from
Motala on the eastern shore of Lake Vättern, through Lakes Boren and
Roxen to the hamlet of Mem at the head of the long Baltic fjord of
Slätbaken. Here there was no second summit, the fall being continuous from Lake Vättern to the Baltic, and the chief engineering
problem was to negotiate the steep slope down into Lake Roxen at
Berg. Here Telford planned a flight of fifteen locks consisting of four
pairs, or double locks, followed by a great staircase of seven which fell
tier by tier down the last steep slope to the level of the lake.

In 1809 a Swedish translation of Telford's report was laid before
Parliament. It was accompanied by a series of beautiful hand-coloured
maps of the proposed canal which had been prepared by von Platen's
assistants from the survey plans and on to which they had pains-
takingly copied Telford's signature. By this time the political climate
of Sweden had changed in a way which favoured the canal project in
every respect but one. The bloodless revolution of 1809 deposed King
Gustav Adolf in favour of his uncle Carl XIII and adopted a new constitution which von Platen helped to frame. In 1810, Bernadotte, one
of Napoleon's marshals, was elected Crown Prince and quickly became

the ruler of the country in all but name. As the Prince looked upon von Platen as his chief adviser, the latter's hand was very much strengthened. State financial support for the canal scheme had been approved by the Government in 1809 and now at last the Canal Company was launched with great éclat, the capital being quickly oversubscribed. The project seems to have been treated as a military operation and within four days of the formation of the Company a detachment of the Swedish Grenadiers began work on the eastern line at Motala under the direction of Samuel Bagge.

So far so good, but although von Platen had at his disposal a prodigious labour force – altogether 60,000 soldiers and seamen were eventually employed on the canal – it was totally inexperienced in the art of canal construction and he badly needed a leavening of skilled men from England. He also needed specimens of English building equipment and materials to serve as patterns for his countrymen to copy. This was where the difficulty lay, for the new régime had shown itself more favourable to Napoleon than its predecessor, with the consequence that diplomatic relations with England had become more strained. However, by sheer perseverance the Count eventually got what he wanted.

Von Platen was an inveterate and tireless correspondent. The Harwich packet *Diana* still lay in Gothenburg roads awaiting a change of wind when Telford received the first long letter from the friend he had only just left. The Count was most solicitous: 'A few hours after your departure I found how well it would have been to see You once more . . .', he wrote, '. . . I wisched You would, if the wind is settled strong westerly, come once more ashore as I am detained here in all case till Wednesday evening; my reasons are 1st that fearing Your provisions of wine and liquors will soon fall short and not knowing if the Capten has provided for of having other, I wish You would provide for a new stock from shore . . .' Evidently Telford had presented his friend with a theodolite just before he left, for another of the Count's reasons for wishing him back was the following: 'At looking at the levelling instrument I found I had better take advices of you about it than standing talking nonsens last evening up stairs, but as this was not the case Self interest should bring me even for this reason to wish you back for a little while.' 'Lastly', von Platen concluded, 'You see mankind are always so foolish either to run away from or run after each other.

1, 2. Two Telford portraits: The painting by Sir Henry Raeburn is romanticised whereas the cameo by an unknown artist is probably the most characteristic likeness we have of Telford in middle life.

CHURCH of ST. MARY MAGDALEN in the TOWN of BRIDGENORTH in the COUNTY of SALOP.

3. Telford the architect: Elevation of the church of St. Mary Magdalen, Bridgnorth.

4. The stream in the sky: Pont Cysyllte Aqueduct. The lithograph is by G. Pickering and is in the Victoria and Albert Museum.

5. Cartland Crags Bridge, from an oil painting in the Library of the Institution of Civil Engineers.

6. Telford's first Scottish bridge, Tongueland, River Dee.

7, 8. The Menai and Conway Suspension Bridges. Two illustrations from William Provis's book on the bridges drawn by G. Arnold and engraved by R. G. Reeve.

9, 10. The Menai and Conway Suspension Bridges today.

11, 12. Iron mastery.
Above left. Detail of Spandrel, Waterloo Bridge, Bettws-y-Coed.
Above. Two of the suspension chains, Menai Bridge.

13. Over Bridge, river Severn, Gloucester, showing the unusual design of the arch.

14. The opening of the Gotha Canal: The Royal Yacht *Esplendian* entering the sea lock at Mem, 26 September 1832. From a painting in the possession of the Gotha Canal Company at Motala.

15. Unachieved masterpiece: Telford's magnificent design for a new London Bridge, span 600 feet.

16. Count Baltzar von Platen, from a portrait in the possession of the Gotha Canal Company.

17. Thomas Telford in the last year of his life. From a drawing by William Brockendon in the National Portrait Gallery.

Now, as in the beginning running away would not do for our business, we soon found since that it would not do neither for our affections; and thats the reason why I now find myself always in the habitude of running or looking after You.'

It was a quaint letter to address to a middle-aged Scot after a mere six weeks' acquaintance and we can only speculate upon Telford's reaction to it. For the weight of his engineering business was now so overwhelming that with the best will in the world he could not keep pace with the Count's letters and his less frequent replies, though friendly enough, were generally brief and to the point. This, however, in no way inhibited von Platen; the spate of his letters continued unabated. 'It would be a very tedious business', he wrote, 'if in our correspondence I always should wait upon your answer to a foregoing lettre before I made a new one.' He goes on to remind Telford about the specimen tools which he wants: 'I forgot in my last to speak about the tools you promised to procure; you have mentioned your having spoken to our Envoye about it and your having advised him what to do, but you know my opinion about such Ministerial transactions about things wich are thougt only trifles by people who only think of consequence the killing of some thousand men.'

Sheer pertinacity won the day and despite the difficulties of communication between the two countries von Platen got his tools: three railway wagons of different types, three wheelbarrows, picks and shovels, and six lengths of cast-iron tram rails together with spike nails and switch parts. Along with them went a large bundle of drawings addressed for the attention of Samuel Bagge, that worthy having written to assure Telford that he would be 'your Davidson in Sweden'. There were drawings of lock gates, lock walls and stop gates, swinging and lifting bridges, and of an inclined plane as installed on the Caledonian Canal construction tramway. There was even a drawing of the new canal office building at Ellesmere and it would be interesting to know whether a replica of it exists somewhere in Sweden.

Soon after the plans of the Gotha Canal had been completed and submitted to the Swedish Parliament, Telford was astonished to receive a communication from the British Consul in Gothenburg informing him that the King had created him a Knight of the Royal Order of Vasa. Telford wished to decline the honour on the grounds that he was not resident in Sweden and that in any case such a distinction should be deferred until something more tangible than paper plans had been achieved. However, he found that the letters patent had already arrived in England so that he could not do other than accept. When William Chambers, who was born in Gothenburg the son of a Scottish merchant there, received a similar honour for his architectural achievements he secured recognition for his Swedish knighthood in England. Telford did not follow his example. 'It is more by the performance of useful works than the enjoyment of splendid orders I wish my name to be known to present or future generations,' he wrote. So plain 'Mr. Telford' he remained, except on letters from Sweden which were always punctiliously addressed to 'Sir Thomas Telford'.

The first practical work undertaken on the canal was straightforward digging and this went smoothly enough under Samuel Bagge's direction, although the bitter Swedish winters brought all activity to an end. 'It is a pitty you stopt not here for a winter for to see us all converted to snow and ice,' wrote von Platen in February 1809, but perhaps Telford was not sorry to miss this spectacle. The Highlands of Scotland could be quite rigorous enough for any man. By the summer of 1811, 5,000 men were at work on the canal but by now the commencement of work on the locks and other masonry was overdue and for this the Count sorely needed those 'Superintenders and Capital workmans' from England for whom he repeatedly urged. Fortunately for him relations between Sweden and England improved once more as a result of Napoleon's disastrous campaign in Russia, and in 1813 Telford was able to pay his second and final visit to Sweden, taking with him two assistants, John Wilson and John Simpson's son, James. Apparently he found that work had been started sporadically all along the line, for he urged von Platen to concentrate upon the completion of certain lake to lake sections so that these could be opened for traffic and thus earn revenue for the Company.

Telford's business in England and Scotland was far too pressing for him to remain long in Sweden, but he left Wilson and Simpson behind him and in the following year they were joined by six more men. These were: Alexander McKenzie of Strathpeffer, Thomas Carlisle of Fort William, William Urquhart and Alexander Fraser of Inverness, David Lyon of Aberdeen and James Smith of Peterhead. Evidently Telford had a marked preference for Scotsmen, though the contingent was presently joined by a single Yorkshireman named Ashworth.

Meanwhile two young Swedish engineers, Edström and Lagerheim, were sent over to England by von Platen for a nine months' period of training under Telford at the Company's expense. Surviving among Telford's papers are the bills for purchases made by these two young men, charged to Telford's accounts and, presumably, charged by him to the Gotha Canal Company. They bought themselves four shirts apiece and a pair of 'superfine blue pantaloons' from Alexander Macdonald of Inverness, while Telford's bookseller, Mr. Joseph Taylor of 59, High Holborn, must have been delighted by their order for no less than £81 worth of books. Most of the long list of titles is relevant enough, but if von Platen ever checked the account he must have been puzzled to know what *Clarke on Pregnancy* and *Burns on Abortions* had to do with civil engineering, and may have speculated uneasily upon the activities of his young countrymen abroad.

The arrival of Telford's Scottish recruits in Sweden was timely, for three months later while von Platen was away on a diplomatic mission to Pomerania he received the news of a major disaster. 'The very worst canal news that possibly could happen has reached me here,' he wrote to Telford. '... Major Bagge in the middle of the Wattern has oversett in a gale of wind in that small boat in wich you went together to the Freestone Quarrie and has been drowned with all his company! Had with him all counting books for the West side and plans to be produced at the General Meeting next month. ... Poor Bagge ... he thought himself a seaman when realy he was one for fine weather; but as for the Canal business he was more to me in all respects than ever can be filled up by any boddy in Sweden. ...' 'I can readily conceive your distress at the loss of Major Bagge,' Telford replied. 'In the language of our Shakespeare "You could have better spared a better man", but you must make men for your purposes as I have done ...'

It was not easy, however, to make good such a loss. With the Count away there was no sure hand at the head of canal affairs; the staff were soon at loggerheads and the progress of the work suffered in consequence. The disaster had occurred in August and in the following October von Platen was writing: 'I want to come home to keep the rest of the people a little in order as by the death of Bagge they think them all Chiefs.' The fact was that the Scottish contingent was not settling down too well. They disliked the military atmosphere in which the work was conducted and this, combined with their own manifestly superior technical abilities, tended to make them exclusive, resentful and arrogant. Their feeling is summed up by James Simpson when he wrote to his father: 'Heare they are very lofty or wishing to be so. I have always found a good friend in Baron Platen and for the others they are very ignorant in regard to Canaling. We are ruled hear [*sic*] by Majors – Captains – Lieuts, &c. and no person heare is much respected without he belongs either to the Navy or the Armies.'

It soon transpired, however, that young Simpson was the black sheep in Telford's flock. He had asked the Count for reports on his men and soon after the latter's return to the canal works he said of Simpson: 'Young and inexperienced of the world, he may when coming in a foreing country for to teach other men to work have thougt himself above all controul as well to his working as to his living.' Evidently Simpson did not mend his ways, for twelve months later von Platen was less disposed to find excuses for his conduct. 'I think myself obliged to tell you', he wrote, 'that I consider him a man not much to be depended on in his business as he allways will make more for a shining appearance than for true solidity. He has a deal of pride, thinks himself superior to his countrymen and endeavours by no means to sustain this superiority by a strict conduct but drinks heartily along with them at their expense as he is a good econome of his own, whilst his turn of day passes very soberly. This was the case in the beginning but his Countrymen being sober and quiet people separated from him after a fighting or two . . .' But von Platen was full of praise for Wilson's work on the chambers of the great flight of locks at Berg on the eastern line, while his brief pen picture of the stolid Yorkshireman, Ashworth, is delightfully characteristic: 'As for Mr. Ashworth,' he wrote, 'he continues in his old way to the great contentment of everybody, his religious zeal keeping within doors; certainly he tells me sometimes with a plentifull portion of Yorkshire-talking that he thinks he should have a little more pay . . .'

Matters came to a head when the Scottish contingent was returning home for the winter of 1815–16. John Wilson had expressed a wish to see the Trollhätte Canal, so von Platen had arranged for his brother-in-law, Ekman, to conduct the party there and then to see them aboard the packet at Gothenburg on which the Company had booked their passages to Scotland. All went well until they reached the quayside, when a first-class row broke out. Acting as spokesman for the party,

John Wilson refused the booked accommodation, insisted that they must travel first class, roundly abused the unfortunate Ekman and concluded by saying that nothing would induce him to set foot in Sweden again. Von Platen was greatly upset by the news of this imbroglio and on Christmas eve he dispatched a detailed account of the affair to Telford. Notwithstanding the fact that Wilson was the ringleader he was convinced that James Simpson was at the root of the mischief. 'I never patiently suffered any injustice of a king, consequently not of Mr. Wilson,' he concluded, 'but I forgive him easely considering what opinion I have of him. ... Simpson egged him on; he has a decided bend to intrigue, desimmulation, laziness and pride.'

This, needless to say, was the end of James Simpson's connection with the Gotha Canal greatly, no doubt, to the distress of his father whose death followed so soon after. 'We have been disappointed in Simpson. I have, of course, done with him – he is unfortunately undeserving,' was Telford's terse comment. It seems to have been no less than just to judge from the singularly unattractive portrait of the man which emerged from von Platen's letters. But where the rest of his men were concerned, Telford contrived tactfully to suggest that von Platen might be a little more tolerant in future and that on occasion a certain shortsightedness could make life easier for all concerned. 'You are well aware', he wrote, 'that much trouble attends governing mankind. I have my share of it (in managing Overseers) and although when matters are unfortunately brought forward I never give way, yet I am often glad to seem not to know lesser improprieties unless they are forced on me.' He went on to suggest that the men should be offered increased wages for the next season. In the event, with the exception of Simpson, all the men returned to Sweden in the spring of 1817.

Meanwhile, as in the case of the Caledonian, the Gotha Canal, though still far from complete, had already exceeded the original estimates both in capital cost and time. Poor von Platen's first appeal for an additional grant from the Government in 1815 had been met by fierce criticism and bitter opposition. 'What a damnd noise, what an outcry!' wrote the exasperated Count. 'All the sence and nonsence of the whole country at once in motion! Till beginning of August I have mostly alone been standing in Butt in the House of Nobles. Its Schoking! ... Dam them all, they have embittered a whole summer to me and I have noon to spare any more.' Nevertheless, the indomitable man succeeded in getting a grant of 300,000 Riksdahler a year for six years. In 1817 he had to face another storm over the canal in the Swedish Diet and Telford wrote consolingly: 'I am sure your good sense and manly firmness will always be prefered to despite the opposition and grumbling of those who, unable and unwilling to conduct any great and useful work, are the first to bestow unjust blame upon those who manage them. After being thoroughly convinced of the solid merit of a public work I remain equally indifferent to the praise or blame of those unqualified to judge.'

It was in May 1817 that Telford sent out James Thomson of Glasgow for the purpose of superintending the erection of an iron foundry beside the new canal at Motala where lock gate frames and other ironwork could be produced. Unfortunately, like James Simpson, Thomson was not a success in Sweden and he was back within the year. '. . . More modesty in his reserve and manners would more promote his interest and respectability' wrote Telford, and another Scot, Daniel Fraser, was sent out in his stead. Fraser succeeded where his predecessor had failed. He made his home in Motala and eventually died there, while the works he helped to found eventually grew into the Motala Verkstad, one of the most important industrial concerns in the country. It was largely due to the fertilising agency of this new ironworks that of the four new towns which von Platen planned to build beside his great waterway, Motala was the only one to take root and flourish.

In the autumn of 1819 a pair of iron lock gates weighing forty-eight tons were cast for the Gotha Canal Company at Hazledine & Thomson's foundry at Broseley, Shropshire and sent down the Severn for shipment at Bristol along with other, smaller specimens of ironwork such as heel posts and paddle frames. Six months later von Platen was able to tell Telford that this work of the Shropshire ironmasters had been successfully 'imitated' at Motala.

There was a second heavy shipment of equipment from England in April 1822, this time from the Thames. Following the success of the two steam dredging machines on the Caledonian Canal, von Platen, on Telford's recommendation, ordered a similar machine for the Gotha from the Bryan Donkin Company of Bermondsey. In this case some details have survived which are lacking in the case of the Caledonian machines. It was designed to work to a depth of twelve feet. The dredger ladder was of wood and carried twenty-two wrought-iron buckets 'steeled at the mouth'. It was driven by an 8 h.p. steam engine supplied with steam from a wrought-iron tubular boiler. The price delivered on shipboard was £1,500. At the same time, Bryan Donkin supplied at a cost of £1,800 the machinery for two small paddle steamboats. This consisted of two 10 h.p. engines and two tubular boilers complete with the necessary paddle wheels 'with iron arms and floats, shafts with coupling links and plummer blocks with brasses all fitted complete'. All this machinery was installed in hulls built at Motala.

Though by no means so tireless a correspondent, Telford frequently sent von Platen reports of the progress of his works in England and Scotland and pressed the Count to visit him so that he could see them for himself. Both men were beginning to realise that advancing years and the ceaseless pressure of work from which they never for a moment relaxed were beginning to tell and that the sands of life were now fast running out. 'If you do not come this year or next', Telford urged at the beginning of 1822, 'there is a risk of the Game being over for us both. We play with so much eagerness – it cannot last long . . .' So, leaving the Gotha Canal works in the capable hands of young Lagerheim, who was now his chief engineer, von Platen came to England at last and spent some delightful weeks touring with his friend. His gratitude for this brief respite and his sorrow at parting are touching. 'My dear friend I shall leave this shore perhaps forever but certainly these last 5 weeks of stay in England will be remembered for ever by me . . .' he wrote while his ship still lay in Harwich roads, and when he reached home: 'I feel me happy in sitting down for to converse an hour with him who along with my wife is my best friend and foremost support in this world . . . to whom I owe wholly one of the most interesting and the most instructive and useful periods of my life; who only makes me regret to be perhaps so near the term of my activities.' They never met again.

Von Platen's words were prophetic; perhaps he already suspected that he was suffering from a mortal disease. But he enjoyed his moment of triumph in the September following his visit when the difficult part of the western line of the canal from Lake Vänern over the summit to Lake Viken was at last completed and he was able to send Telford a glowing account of the opening ceremony. 'Dear Friend!' he wrote from Mariestad, 'The 23rd of this Month the Great Ceremonie of the kings passage from the Wenner to the Wikken on the Gota Canal took place with full success. A particular Wessel was fitted out for this purpose and was accompanied by to large Gunboats schooner rigged and several other Wessels. The water in the canal was between 8 and 9 feet and under an immense crowd of people parading Regements with their bands and firing of different batteries the Scene was realy grand. The only thing missed not only by me but by many other was the presence of You dear friend; but I know You are not found of Ceremonys. . . .'

This tangible evidence of success silenced the canal critics and so made von Platen's position easier – but only for a short time. In the next working season of 1823 when it was hoped to complete the western line to Lake Vättern at Rödesund great difficulties were encountered, the worst in the whole weary, protracted struggle to finish the canal. Just as the Caledonian Canal suffered after its premature opening due to an insufficient depth of water in Loch Oich, so now it was found that the narrow, winding passage through Lake Viken was extremely shallow and hazardous. But whereas Donkin's dredging machine gradually ate its way through the silt shoals which the Garry washed into Loch Oich, it was powerless against the jagged fangs of rock that lurked beneath the surface of Lake Viken. There was only one way – a costly and infinitely laborious one – of dealing with these rocks. This was to expose them by means of coffer-dams so that they could be hacked and blasted down to a safe level. The disastrous failure of a coffer-dam at Forsvik would have defeated von Platen's hopes of clearing a passage through the lake before the winter had not one of the toughest regiments in Sweden been assigned to the task under his personal command. When the onset of the northern winter brought all other work on the canal to a standstill, these picked men, spurred on by the Count's unquenchable spirit, worked on under conditions of incredible hardship until the job was done. 'Some individuals having uttered a wish to see their home, [I declared] that neither they nor I would see our home before we had finished the task!' he wrote, 'and after this open declaration, with some dexterity, daily and hourly attendance of mine, and some brandy and ale, I never heard a murmur of 517 men quartered on the Rocks of Wiken in almost open sheds and often working in water to the knees tho' the ice in other places was strongth to bear a man! Happily I was in good health tho' I was with them every hour and at the 25 of Octobre the bataillon left the place without a single sick under laughing and singing.'

So this major hazard was overcome, but still the work dragged on from year to year and still von Platen had to fight his endless battles against violent opposition in order to get the necessary funds. By 1825, 100 miles of navigation were open, the whole western line and the eastern line as far east as Lake Roxen. He had hoped that 1829 would see the canal completed, but it was not to be and in that year his appointment as Viceroy in Christiania took him away from his beloved canal. As he related bitterly to Telford in a letter from Christiania in May, opposition to the canal had gathered strength in his absence and the long hoped for completion had receded even further away. 'I see clearly by all these protractions the coming into the Baltic will not be affected before 1831 tho' we had intended it 1830. As for the rest the Navigation is brought in sight of Söderköping: I have even this winter victoriously shown it to be the foundation of Scandinavian independence in peace as in war when I spoke in the House of Nobles. And so I have beforehands settled my mind, may the issue be what it will. I have done my duty tho' with little gratitude.'

This was his last word to Telford. Although only one short link from Söderköping to Mem on the Baltic had still to be forged before the Count's dream of a waterway from sea to sea was realised, he did not live to savour that realisation. For Admiral Count Baltzar von Platen died of cancer in Christiania in December 1829 at the age of sixty-four and it was not until 26 September 1832 that his object was finally achieved. On that day the King, Queen and Crown Prince of Sweden in the royal yacht *Esplendian* sailed from the Baltic into the new sea lock at Mem and so declared open the great inland route to the North Sea. This was the consummation of twenty-two years of endeavour represented by a staggering total of 100 million man hours and a cost over five times that of the original estimate.[1] Heedless of the cheering and the thunder of cannon which celebrated this final victory, von Platen slept peacefully beneath a huge monolith of granite in the grave he had chosen for himself on the bank of his canal between

1. Like the Caledonian, the Gotha was built during a period of rapid economic inflation and in both cases this partly, though by no means wholly, accounts for the excessive capital outlay.

Motala and Lake Boren. Lagerheim, his chief engineer, also lies there beside the master he served so faithfully.

The Gotha Canal is first and foremost a monument to that great patriot, the Count von Platen, but the marks of Telford's hand are plain and clear. In many respects, particularly in details of lock design and ironwork, similarities between the Gotha and Caledonian canals betray a common authorship. The Gotha lock gate capstans, for example, with their wooden capstan bars up-tilted in iron sockets, are obviously based on the Caledonian design, although in the case of the Gotha they swing the gates by means of pinion gears and long horizontal racks instead of chains. Like the Caledonian, the Gotha never fulfilled the high hopes of its promoters for similar reasons. The increase in the size of ships, the coming of steam power at sea and on rails and the abolition in 1857 of the heavy tolls in the Sound between Sweden and Denmark all combined to limit its success.

At the time of its building the Gotha Canal exceeded in size the older Trollhätte Canal between Gothenburg and Lake Vänern, but today the position has been reversed for whereas the Trollhätte has been greatly enlarged and completely modernised the Gotha remains as it was when von Platen and Telford built it. Nevertheless, unlike the Caledonian, it still carries considerable commercial traffic notwithstanding the fact that it is icebound during the winter months. It also passes much pleasure traffic in summer, including a regular service of passenger steamers which ply between Gothenburg and Stockholm. A sad contrast this to Scotland where no ships steam along Telford's superb waterway through the Great Glen and visitors must perforce roar down the Glen road in buses or cars, seeing little or nothing of its beauties.

When Telford heard the news of the completion of the Gotha Canal he was seventy-five and even his splendid constitution was failing him. But in the years that had gone by since the day he had received that first letter from Sweden at Inverness he had accomplished more than enough to fill a long lifetime. Not only had he seen his great works in the Highlands carried to completion, but, in stone, in water and in iron, he had left his indelible mark upon the Midland shires of England from the Black Country to the Mersey, from London to the Welsh border and from thence to the furthest west of Wales.

8

THE ROAD TO HOLYHEAD

Dwy flynedd, cyn aflonydd
Pont ar Fenai a fydd.

A CONTEMPORARY print, dated about 1820, shows Telford on a tour
of inspection in the Highlands. He is driving himself in a smart post-
chaise drawn by a high-stepping grey, sitting erect, a large, impressive
figure in a caped greatcoat, black top hat and long driving gauntlets.
The scene is one of rugged grandeur: mountains and beetling granite
crags; a clump of pine trees on a rocky knoll and in the middle distance
a broad river spanned by a three-arch bridge of stone. Most probably
this scene existed only in the mind's eye of the unknown artist, but it
is none the worse for that, for it is against just such a background as
this that imagination will always picture Telford. One might almost
suspect that the scenes of his greatest engineering achievements were not
determined by any mundane considerations of utility but were chosen
with an infallible eye for dramatic effect. Certainly that cult of the
picturesque which was at this time transforming at vast expense so
many acres of English parkland never produced results comparable with
those which Telford achieved fortuitously. The Vale of Llangollen
gives the aqueduct of Pont Cysyllte its splendid frame; the cloud-
capped crags and screes of Britain's loftiest mountain look down on
Neptune's Staircase at Banavie. And now the scene changes back again
to Wales, where Telford was to drive his Holyhead road through the
Pass of Nant Ffrancon and build the most famous of all his bridges
against a magnificent background of mountains – Snowdon, the
Glyders, Carnedd Llewelyn.

At the same time that Beaufoy and Dempster were urging the
Government to help the Highlands of Scotland, a small group of Irish
M.P.s had been pressing the need for improved communications
between London and Dublin. It was as a result of their efforts that John
Rennie and Captain Joseph Huddart were appointed in 1801 to report

on the most suitable route and the most favourable sites for cross-channel packet harbours. They recommended Holyhead and Howth and after six years had lapsed Parliament voted a sum of money for the construction of harbour works at these two places. Nothing was done in the way of road improvement, however; indeed, John Rennie held the view, which was quite common at that time, that road-making was an activity beneath the notice of a civil engineer.

No less than seventeen Turnpike Trusts were responsible for the road between London and Shrewsbury, and there were seven more between Shrewsbury and Holyhead. The state of a particular stretch of road depended upon the finances of the Trust responsible and these depended in turn upon the population and traffic in the locality. The effect of this was that the further the traveller went westwards the worse the road became. The London mail coaches could not operate west of Shrewsbury, while in the island of Anglesey the road was so bad that it scarcely deserved the name. A contemporary called it 'a miserable tract, composed of a succession of circuitous and craggy inequalities'. In 1808 the Post Office made an attempt to extend the mail coach service to Holyhead, but it was completely abortive. Even for a riding post the route was perilous as was shown when, within one week, three post horses fell and broke their legs. The Postmaster-General attempted to force the Welsh Trusts to comply with their legal obligation to keep the road in good repair, but it was obvious that this was quite beyond their means and that nothing effective would ever be done without financial help from the Government.

It was at this juncture that, at the instigation of John Foster, the Chancellor of the Irish Exchequer, Telford was instructed to examine the route between Shrewsbury and Holyhead and to make a report. Unlike Rennie, he had no inhibitions about road building. Starting out from Shrewsbury, he first explored two alternative and more direct routes across the Berwyns. He gives us no details of these, but they would almost certainly have followed the valleys of the Tanat and the Ceiriog. He decided that they would involve too severe gradients and he therefore recommended the improvement of the accepted, though more circuitous route along the valley of the Dee through Llangollen to Corwen and from thence up the course of the tributary Ceirw, and down into the Conway valley. From Bettws-y-Coed the mail route took an extremely devious course down the east side of the Conway and then along the coast through Penmaenmawr to Bangor. But, said

Telford in his report, the line of a suggested coach road had already been marked out through Capel Curig, Lake Ogwen and the Nant Ffrancon Pass. Who had marked it out he does not say, perhaps it was Huddart, but Telford thought it practicable, and it was certainly much shorter. In Anglesey he recommended an entirely new road. The old road, he said, was not more than 12 to 13 feet wide, as steep as 1 in $6\frac{1}{2}$ and 'passed along the edge of unprotected precipices'.

A Parliamentary Committee issued a report based on Telford's findings and recommendations in May 1811, but unfortunately this coincided with the abolition of the office of Chancellor of the Irish Exchequer and the retirement, with a peerage, of John Foster. Robbed of its moving spirit, the campaign languished and, apart from a few bridge works which were put in hand, the report gathered dust for five years before another energetic champion came forward in the person of Sir Henry Parnell, the Member for Queen's County. A man of great energy, enthusiasm and determination who would not take no for an answer, Parnell was undoubtedly the von Platen of the Holyhead Road. It was due to his efforts that a Holyhead Road Commission was set up in 1815 and Telford was instructed to carry out a detailed survey of the whole route from London to Holyhead.

Telford set to work immediately by appointing a team of three to carry out the survey under his direction. They were, in order of seniority, William Provis, John Sinclair and Robert Sproat. Provis was that promising young assistant of Telford's whom we have already met; Sinclair and Sproat were taken off the Highland roads scheme where they had been working as assistant inspectors. This survey alone was no small undertaking, and it occupied the trio continuously from early September 1815 to the end of March 1817. As an example of the cost of a work of this kind at that time, Telford's account to the Commissioners may be quoted. It was made up as follows:

Telford's Fee	..	..	892	10	0
Expenses	..	..	426	4	2
Paid Assts	..	..	832	7	$11\frac{1}{2}$
Total		..	2151	2	$1\frac{1}{2}$

This figure, of course, would include the cost of preparing plans and sections covering the whole route, together with a detailed report by Telford explaining his recommendations.

On the road between London and Shrewsbury Telford's proposals are reminiscent of a typical programme of road improvement such as is carried out today. A series of small maps which accompanied the report show gradients eased and sharp corners or narrow sections in villages avoided by short new by-passes. The first important improvement at the London end was the Highgate Archway cutting. Others recommended were at Barnet and South Mimms, at Fenny and Old Stratford, while hills were to be eased at Hockcliffe, Brickhill, Fosters Booth, Braunston, Meriden, Cosford, Priors Lee and Ketley. Between Shrewsbury and Holyhead work of a far more radical nature was required. Indeed here it was not so much a matter of road improvement as of road making, and Telford recommended that the Commissioners should buy out the seven Turnpike Trusts involved and become directly responsible for the reconstruction and subsequent maintenance. The survey also covered the North Wales coast road from Chester to Bangor, as this formed part of the mail route from Liverpool and Manchester to Holyhead. The magnitude of the works which Telford proposed may be judged by the fact that he estimated that on the section between Chirk and Holyhead alone an outlay of £53,000 per annum for five years would be required.

Work on the Holyhead road went on for fifteen years, although long before it was finally completed coaches were able to run through to Holyhead. The great expenditure both in money and time was partly due to the high cost of the road-building technique which Telford had perfected in the Highlands. It is because of this high cost that Macadam is better known as a road maker than Telford. Macadam was content to level the ground and lay his road metal upon it without any massive foundation, claiming that a certain resilience in the surface was an advantage and easier on horses. Whether this was true or not, the speed, the ease and the cheapness of Macadam's method led to his widespread employment by Turnpike Trusts throughout the country, whereas the roads to Holyhead from Shrewsbury and Chester are the only noteworthy examples of Telford's road-making south of the Border. But what roads they were! Telford built them in the Roman fashion and it was his boast that, like theirs, they would last for centuries.

Robert Southey, in his diary, tells us how Telford made his roads, and as his account is obviously based on the engineer's own explanation it is virtually first hand. 'The Plan upon which he proceeds in road-making is this,' Southey writes, 'first to level and drain; then, like the Romans, to lay a solid pavement of large stones, the round or broad end downwards, as close as they can be set; the points are then broken off, and a layer of stones broken to about the size of walnuts, laid over them, so that the whole are bound together; over all a little gravel if it be at hand, but this is not essential.' Telford laid great emphasis on the proper grading of the stone, and this graded stone was stored in kists or bins by the roadside for surfacing and for subsequent repairs.

Elsewhere in his Highland diary, Southey describes the care taken in constructing the road along hill slopes. 'If the hill be cut away,' he writes, 'it is walled a few feet up, then sloped, and the slope turfed; if there be no slope, a shelf must be left, so that no rubbish may come down upon the road. The inclination is toward the hill. The water courses are always under the road, and on the hill-side back drains are cut, which are conducted safely into the water courses by walled descents, like those upon the Mont Cenis road, but of course upon a smaller scale. This road is as nearly perfect as possible. After the foundation has been laid, the workmen are charged to throw out every stone which is bigger than a hen's egg.[1] Every precaution is taken to render the work permanent in all its parts. Thus where a beck coming down from the hills is bridged, the beck itself for some distance above the bridge is walled, to keep it within bounds; the foundations of the bridge are laid two feet below the bottom of the stream, and for farther security, the bottom itself, under the arch, is secured by an inverted arch of stones without mortar.'

The action of iron tyres on the surface of Telford's roads was to break off minute fragments of the top layer of stones and pack this grit between these stones so that a perfectly smooth surface was the result. Rain did not easily penetrate this surface and as the roads were built with a generous camber, the water ran off into the side ditches. Given reasonable maintenance, both Telford and Macadam provided by these methods a smooth and lasting surface which alone made possible the high speeds achieved during the short heyday of the road coach. This form of surface was not, however, proof against the action of the pneumatic tyre which, instead of consolidating it, sucked the grit from between the stones to be turned into mud or choking clouds of dust.

1. The Telford *Atlas* shows, with other road tools, a ring gauge, 2½ inches in diameter, which was presumably to be used for stone grading.

Work was begun on the worst and most dangerous sections west of Llangollen and by 1819 sufficient progress had been made for mail coaches to run through to Bangor with safety. Telford had decided that there must be no gradient steeper than 1 in 20 and this, in such mountainous country, involved more than mere side cutting. In places deep cuttings had to be blasted through rock and in others considerable embankments had to be formed or high retaining walls built. Proceeding westwards, the first of these difficult sections was between Berwyn and Glyn Dyfrdwy only a few miles west of Llangollen and there was another on the summit between the valleys of the Ceirw and the Conway. At Bettws-y-Coed the road crossed the Conway by the most notable of the bridges which were put in hand before 1817. This was a single span of iron which, needless to say, was cast by William Hazledine at Plas Kynaston and erected by William Stuttle. It is an extravagant masterpiece of the iron founder's art. The webs of the cast-iron girders consist of the inscription: 'This arch was constructed in the same year the battle of Waterloo was fought' in iron lettering, and the spandrels are filled by elaborate devices incorporating the emblems of the four nations: the rose, the leek, the thistle and the shamrock. It is a display of exuberance rare in Telford's work and unfortunately its proper appreciation involves clambering down to the riverside, so that it is wasted upon the thousands of motorists who hurry over the bridge every year.

The long winding ascent from Bettws-y-Coed to Capel Curig is a spendid piece of road engineering, but undoubtedly the highlight of the whole route is the section which follows – the road past Lake Ogwen and through the pass of Nant Ffrancon. Here, by masterly engineering in a setting of desolate grandeur which rivals anything on his Highland roads, Telford carried his road over the high summit on a ruling gradient of 1 in 22. It remains to this day by far the most easily graded of all the roads which traverse the North Wales massif. As the motorist speeds easily over the pass in top gear he pays tribute to the power which twentieth-century automobile engineering has built into his car, but seldom or never gives credit to the men who surveyed his road 140 years ago. At the head of the pass a cutting blasted through the rock and protected from rock falls by revetments, an embankment and a stone arch carries the road dramatically across the chasm through which the river Ogwen thunders down towards the floor of the valley.

Notwithstanding the dramatic splendours of the Nant Ffrancon pass, however, the most costly section of the whole road, excluding major bridges, was not here but a few miles from Holyhead where a narrow, silted strait separates Holy Island from the mainland of Anglesey. At the village of Valley in Anglesey the old road veered sharply to the south-west to cross this strait at its narrowest point by what is called the Four Mile Bridge. This involved a considerable detour and Telford decided to continue this last lap of his new road on a straight course, a decision which involved building a great embankment across the Stanley Sands reminiscent of the Mound at Dornoch Firth. It was 1,300 yards long, 16 feet high, 114 feet wide at base and 34 feet at top with sides protected against storm erosion by rubble walling, yet the contractors, Messrs. Gill & Hodges, completed it in one year. It was opened in 1823.

So much for the Holyhead Road itself; there remained the greatest problem of all which, at the time of the 1815 survey, the Road Commissioners still hesitated to confront. This was the crossing of the Menai Straits. In the sixteenth century the Welsh seer, Robin Ddu of Bangor, delivered himself of the prophecy which appears at the head of this chapter. Translated it reads: 'Two years before tumult there will be a bridge over the Menai.' Like all the best prophets, Robin Ddu combined caution with a splendid dramatic sense. His words have a fine gnostic ring about them, yet even in the sixteenth century the need for a bridge was so obvious that its eventual construction was a fairly safe prediction. And as the affairs of men are such that not a year of recorded history has gone by without some major tumult breaking out somewhere or other in the world, the seer was taking no chances on the date question. However, those who honour such prophets would say that Robin Ddu was very far-sighted, for over two centuries were to pass before his bridge became a reality.

Within the lifetime of Robin Ddu the Menai Ferry rights were leased by Queen Elizabeth I to a certain John Williams, and they descended in his family down to the time with which we are here concerned when an average of 13,000 travellers a year were using this Anglo–Irish route. This traffic yielded John Williams's descendant,

Miss Williams of Plas Isa, the not inconsiderable income of between £800 and £900 per annum. When the bridge was eventually built she was awarded at Assizes £26,557 compensation payable by the Holyhead Road Commissioners for the loss of her rights, a nice little nest egg for Sir David Erskine, Bart., whom she shortly afterwards married. The income which the Williams family had for so long enjoyed was derived from the discomfort of innumerable travellers. Notwithstanding all the hazards of the old road through the mountains, it was the crossing of the Menai which travellers most dreaded. The passage was always attended with discomfort and delay, if not with danger. Only in fair weather and calm seas, conditions which occurred seldom in winter, could coaches or carriages be conveyed across. Drovers were accustomed to swim their herds across at low water using the ancient cattle causeway which stretched from the Anglesey shore as far as the rock known as Ynys-y-Moch (Pig Island) and cattle were not infrequently lost in the process.

Schemes for eliminating these difficulties and dangers by means of a bridge had often been debated. John Golborne produced in 1783 an optimistic proposal for a crossing at the Swilly Rocks which lie some distance south-west of Pig Island. This was to consist of an embankment built out to the rocks and a large lock for the passage of shipping on the Caernarvon side which the road would cross by a drawbridge. In January of the following year William Jessop reported in favour of a wooden bridge at Cadnant and in this case a Parliamentary Bill was promoted though the scheme came to nothing. Next, Rennie examined the Straits in 1801 and expressed himself in favour of a three-arch iron bridge at the Swilly Rocks. The only possible alternative, he thought, was the Pig Island site, but this would require a single arch of so great a span that he did not think it a practicable proposition.

The difficulty which ruled out all these schemes was the insistence of the Admiralty that navigation through the straits must not be obstructed but that vessels of the largest size must pass freely with masts erect. They would not even countenance the temporary obstruction of the channel by the centering which would be required to build an arch bridge. These stipulations appeared to be crippling, for they called for an arch of unprecedented height and span and at the same time forbade engineers to use their accepted method of building it.

When Telford was first consulted about the Holyhead Road in 1810 he followed Rennie's lead by preparing two designs of cast-iron bridges, one of three cast-iron arches of 260 feet span for the Swilly Rock site and the other a single span of 500 feet for Pig Island. To overcome objections to the latter he designed a system of suspended centering for which at that time there was no precedent. He proposed to erect a frame fifty feet high upon each abutment of the bridge. Winch cables would be led over rollers on the tops of these frames and would thus be able to lift the sections of the centering into their places and retain them there. It was really the suspension bridge principle adapted to build a rigid structure. Although it was never employed by Telford, this principle was subsequently used by later engineers, not merely for centering but for the erection of actual bridge spans. For reasons already mentioned, Telford's 1810 proposals came to nothing. In any event his designs allowed insufficient headroom to satisfy the Admiralty.

In 1814, while the question of bridging the Menai still lay in abeyance, Telford was consulted about a scheme for bridging the Mersey at Runcorn. Here he decided that a suspension bridge was the only answer although, with the exception of a few very smallscale footbridges, the principle was at that time quite untried in Britain. Assisted by Bryan Donkin, Peter Barlow and others, he made a most exhaustive series of practical experiments into the tensile strength of malleable iron, using for the purpose a special Bramah hydraulic press at William Brunton's Patent Chain Cable works. As a result of these experiments he designed a suspension bridge of truly heroic proportions. It was to have a centre span of 1,000 feet, seventy feet above high water in the Mersey, and two side spans of 500 feet each. Instead of the link chains used in subsequent bridges, Telford proposed to suspend the bridge platform by means of Brunton's laminated iron cables. His drawings for the bridge show four sets of four cables, each set slung vertically one above the other. Smiles in his biography of Telford describes this Runcorn design as 'very magnificent' and 'perhaps superior even to that of the Menai Suspension Bridge', but, looking at the drawings today with an unprejudiced eye, one is forced to the conclusion that it was as well that the bridge was never built. The design would have offered far too little resistance to lateral wind pressure acting upon so great a span, a difficulty which was subsequently countered by Sir Marc Brunel in his design for the Ile de Bourbon bridge where he used very short suspension rods at the centre of the span and a system of inverted chains to provide additional bracing.

While Telford was pursuing his Runcorn experiments, he appears to have been quite unaware of the fact that Captain Brown, later Sir Samuel Brown, was working on similar lines at the same time. Brown's experiments led him to drop the use of laminated cables in favour of a chain composed of long flat iron links and pins. He constructed the Union Bridge over the Tweed at Norham Ford with a span of 361 feet and later several others, including the Brighton chain pier, using chains of this type. As soon as Telford heard of Brown's work he got in touch with him, the two men pooled their knowledge, and just before the Runcorn scheme was finally abandoned in 1817 Telford was contemplating using chains of Brown's design instead of the cables.

The abandonment of the Runcorn bridge project must have been a great disappointment to Telford, who had been writing most enthusiastically about the scheme to his friend von Platen. Nevertheless the experience and knowledge gained by it was soon to stand him in good stead. In 1818 Sir Henry Parnell at last succeeded in forcing the Holyhead Road Commissioners to face the fact that the Menai must be bridged and Telford was again asked to submit plans. His answer this time was to submit a design for a suspension bridge on the Pig Island site. This provided for a single suspended span of 579 feet with a headway of 100 feet which would satisfy the Admiralty. The two main piers, one standing upon the Caernarvon shore and the other on the island rock, would be extended to form the suspension towers, fifty feet above the roadway and 153 feet above high water in the Straits. These mighty piers, each pierced by arches for the carriage ways, were to be approached by seven lofty masonry arches each of 52 feet 6 inches span, three on the Caernarvon and four on the Anglesey side of the Straits. For the bridge platform Telford adopted the same dimensions as those shown in the Runcorn design, namely two carriageways each 12 feet wide separated by a central footway 6 feet wide, making 30 feet overall. To support it there were to be sixteen chains composed of composite links each consisting of thirty-six bars of iron half an inch square. These figures may seem conservative when compared with the grandiose dimensions of the Runcorn bridge, yet they far exceeded those of any bridge ever built before and Telford's plan provoked almost as much scepticism and ridicule as had greeted Brindley's proposal for an aqueduct over the Irwell at Barton so many years before. Nevertheless, thanks to the energetic championship of Sir Henry Parnell, the Holyhead Road Commissioners approved the design and an initial sum of £20,000 to enable work to begin was voted by virtue of their existing powers.

Carpenters were soon on the site erecting temporary wooden buildings for use as workshops and as accommodation for the men to be employed on the work. This done, blasting operations began on Pig Island to level a foundation for the great western pier. At this juncture, however, three influential local landowners, the Marquis of Anglesey, Owen Williams of Craig-y-Don and Asheton Smith, the owner of the Padarn slate quarries, succeeded in stirring up considerable opposition to the bridge in Caernarvon and work was stopped temporarily while the Commissioners appealed to the Government for special powers. The opposition hoped to win support from the Admiralty, Trinity House and the Menai pilots, but these bodies were satisfied that Telford's high-level bridge would not affect their interests and in 1819 the Commissioners obtained their Act authorising them to proceed with the bridge. Work was then resumed without delay. William Provis was appointed resident engineer, Messrs. Straphen and Hall were awarded the masonry contract and Telford himself went down to Bangor to superintend the start of operations.

A source of good hard grey limestone for the masonry work was found at Penmon on the extreme eastern tip of Anglesey, and this was brought by small coasters down the Strait to quays on both shores.[1] A temporary embankment was built beside the old cattle causeway from the Anglesey shore to Pig Island and on this a railway was laid down to convey the stone for the Pig Island pier. It was here that the first stone was laid without any ceremony by William Provis. It was the middle block of the lowest course of the seaward face of the pier. A rock base for the Caernarvon pier was found on the shore six feet below low-water level. By the end of 1819, 200 men and five coasting vessels were at work.

That the engineer in charge of a great engineering work in a remote place had to cope with difficulties undreamed of today is revealed by a request from the Anglesey magistrates that in future the men at work

1. The landowner, Viscount Warren Bulkeley, received 6*d.* per ton from the Government for all stone quarried on his estate.

on the bridge should be paid in Bank of England notes. This led Telford
to explain the method of payment. On receipt of his certificates of work
done the Land Revenue Office issued drafts which Shrewsbury bankers
exchanged at par for their own notes. If, he said, these bankers were
required to issue Bank of England notes instead of their own they
would only do so at a discount. As an example of these exchange difficulties Telford went on to say that when work had begun on the
Holyhead road in the Bettws-y-Coed area he had arranged for wages
drafts to be exchanged at the Denbigh Bank at Llanrwst (whose notes
were doubtless more acceptable locally) but that the bank had failed and
£70 had been lost as a result.

After nine months' work Messrs. Straphen & Hall threw up their
contract. Telford's answer to this was to recall John Wilson from the
Caledonian and Gotha Canals. 'I see Wilson is not to return,' wrote von
Platen sadly in April 1820. With Matthew Davidson and John Simpson
so lately dead, John Wilson and William Hazledine were now the only
survivors of that splendid team which Telford had first assembled for
the building of Pont Cysyllte more than twenty years before. Ageing
now like their captain but as capable as ever, both men rallied to him
to play vital parts in the drama of the Menai Bridge. No doubt John
Wilson was not sorry to return to Wales. He responded promptly to
Telford's summons, bringing with him the two sons who were now
his active assistants and possibly some members of the Welsh colony
who had followed Telford to the Highlands. In their capable hands the
work on the great piers and approach spans went forward so smoothly
and swiftly that in early March 1821 Telford was able to tell von Platen
in one of his letters that the piers had reached a height of sixty feet and
that he was about to begin the detailed drawings for the suspension
towers above them and for the ironwork. In July 1823 he reported that
the piers were up to roadway height, the towers begun and much of the
ironwork ready for installation. In the same letter he mentions that
work on the similar, though smaller, suspension bridge which he had
designed to carry the Chester road over the Conway was also proceeding rapidly.

In order to provide an immovable anchorage for the great suspension
chains, a series of tunnels, each six feet in diameter, were blasted
through solid rock on the Anglesey side to a depth of about twenty
yards. At this depth the tunnels were united by a single chamber in

which was assembled and secured the massive cast-iron frame to which the chains were to be anchored. This frame was therefore securely buried beneath the mass of rock through which the chain tunnels passed. The latter were cut on an appropriate gradient so that the chains would follow a straight path from their subterranean anchorage to the top of the suspension tower. On the Caernarvon shore the procedure was similar, except that earth had to be tunnelled through for some distance before a suitable rock base for the iron frame was found. Here, therefore, massive walls of masonry took the place of natural rock between the chain tunnels, while the distance from tower top to anchorage had to be greater, thus unavoidably destroying the complete symmetry of the bridge.

It was only when these two anchorages had been positively sited in 1821 that Telford was able to settle the lengths and final details of the chains so that a contract for them could be placed with William Hazledine. The links, each a little over nine feet long, were wrought at Upton Forge which stood close by the Shrewsbury Canal near the village of Upton Magna. They were then sent by canal to Shrewsbury where each link was given a tensile test on a machine of Telford's design specially installed for the purpose in Hazledine's Coleham works. John Provis, William's younger brother, was placed in charge of this operation. From the protracted experiments which he had made, Telford had calculated that each link should be able to withstand a strain of $87\frac{3}{4}$ tons before fracture, but in order that there should be no risk of straining them he stipulated that the Coleham test should be limited to 35 tons. As this was equivalent to a load of 11 tons per square inch of cross section, whereas he had calculated that the maximum loading on the bridge would be equal to only $5\frac{1}{2}$ tons to the square inch, this gave the ample safety margin of 100 per cent. To protect them from the corrosive effect of the salt atmosphere the links were heated, plunged in a bath of linseed oil and then stove dried. To ensure accuracy of fitting a steel master link was made through which the three-inch pin holes in each link were bored on a special machine. Today this process would be called jig drilling.

As batches of the links were completed at Coleham they were sent by canal boat (presumably from the Ellesmere Canal wharf at Weston Lullingfields) to Chester, whence they were shipped by sea to the Menai. At the beginning of November 1822, Telford received a letter from Thomas Rhodes at Fort Augustus. Rhodes reported that his work on the Caledonian Canal ironwork was almost completed and that he would be sailing for Liverpool in about a fortnight. Had Telford any other employment for him? No doubt Rhodes had heard what was afoot in Wales from John Wilson, for his seemingly innocent inquiry was perfectly timed. Telford was then expecting the first delivery of chain links from Coleham and Thomas Rhodes was at once engaged to take charge of chain assembly.

In building the Menai Bridge, Telford had no precedents to guide him and he therefore proceeded with extreme caution, never trusting to theory alone, but checking every move by practical experiment. Thus notwithstanding the Coleham tests, the first length of chain to be assembled was tested in tension by slinging it across the valley of the Cadnant brook which was near the site on the Anglesey side. At the same time Telford was able to calculate the power which would be needed to raise it into position on the bridge. A quarter-size model of one chain was also made and suspended so that the calculated lengths of the vertical suspension rods could be checked by actual measurement. While these tests were being made and the chains assembled, Wilson was pressing on with the building of the suspension towers. This was slower work than the piers below, for apart from their great height Telford had decreed that each stone must be dowelled to its neighbour by iron pegs. By the early spring of 1825, however, all the masonry work had been completed and all the chains assembled and fixed to their underground anchorages. Everything was therefore ready for the most delicate and dramatic operation of all – the slinging of the suspended portions of the great chains between the two towers.

One of Telford's characteristics which impressed his contemporaries was his imperturbability. The dangers, disasters and unexpected difficulties to which works such as he undertook are always subject he invariably confronted with an unruffled calm which seems to have come naturally to him. Alone of all his works, the Menai Bridge disturbed that calm. When he arrived in Bangor towards the end of April 1825 to superintend personally the slinging of the first suspension chain, if he appeared to his assistants as imperturbable as ever it was a composure sustained only by an exercise of will. For he confessed later that for weeks past he had been suffering from anxiety so acute that he could not sleep. Like every engineer who abandons all safe precedent and takes a bold step forward into the dark, Telford had found himself haunted by the ghosts of all the things which could go wrong despite all his care and foresight. It was imperative that the whole operation of raising the sixteen chains (each suspended portion weighing $23\frac{1}{2}$ tons), should proceed swiftly and smoothly, for the Commissioners had obtained Parliamentary powers to close the Strait to navigation while it was carried out.

The morning of 26 April broke brilliantly fine; the air still and the sea calm. Telford therefore determined to raise the first chain. The waters of the Menai were crowded with flag-bedecked boats and every vantage point on both shores was thronged with spectators as Telford's carefully worked out plan of operation was set in motion. The chain had been laid upon a raft 450 feet long and six feet wide which was moored near Treborth Mill on the Caernarvon shore and at 2.30 in the afternoon, about an hour before high water, this was cast off and swung out into the Menai towed by four boats. The crowd watched in silence while it was manoeuvred into position between the two great piers. On the Caernarvon side it was then attached to the landward portion of the chain which had been laid from its anchorage over the top of the suspension tower and down the face of the pier to within a foot or so of high-water mark. On the Anglesey side the chain had been laid from its anchorage to the top of the Pig Island tower, so the next move was to raise the floating end of the chain from the raft to the top of the tower. For this purpose cables hung ready down the face of the Pig Island pier. These passed over blocks on the top of the tower and from thence to capstans mounted on the Anglesey shore which were manned by a force of 150 men. As soon as the end of the chain had been securely grappled the men on the raft sang out 'go along' and a fife band struck up, keeping the time as the capstan gangs swung into action. As the cables tautened the shout went up 'Heave away! Now she comes!' and in a few moments a great cheer sounded from ships and shores as the long raft swung away on the tide and the chain was seen to be hanging free in a low arc over the water.

Slowly, slowly as the men circled their capstans the chain rose higher above the water until, after an hour and a half, its end had been drawn up to the top of the Pig Island tower. Telford had himself climbed to the top of the tower with John Wilson, William Provis and Thomas Rhodes to see the chain united. The crowd craned their necks to watch the small black figures grouped together on the giddy platform 150 feet above the sea. When they saw them lift their hats and wave them in the air they knew that the last link pin had been driven home and another mighty cheer went up which echoed and re-echoed from bank to bank across the Menai. The Welshmen who had laboured to such good purpose at the capstans were at once fortified with a quart of 'Cwrw da' apiece and the general excitement and hilarity was such that three of them, Hugh Davies, a stonemason, John Williams, a carpenter, and William Williams, a labourer, actually walked across the chain to the Caernarvon pier. As the chain was only nine inches wide and the curvature considerable, this was a foolhardy feat worthy of the great Blondin himself. Either they escaped Telford's notice or it was one of those lesser improprieties which, as he had told von Platen, 'he seemed not to know'.

The whole operation had been carried out without a hitch in two hours and twenty minutes from the time the raft had been cast off and it was repeated with the second chain two days later. Thereafter a chain was raised whenever the weather served. The lifting of the sixteenth and last chain on the 9th of July was made an occasion for further ceremony. By this time the whole proceeding had become a matter of well-drilled routine, so that only one hour thirty minutes was taken to raise it. As soon as it was home a military band marched down a temporary wooden platform laid over the other chains and halted at the centre of the span where they played the National Anthem. While they played, the navigation through the Straits was formally re-opened by the steam packet *St. David* of Chester, Captain D. Sarsfield, which, dressed over-all, steamed down the channel beneath.

While all this was going on at Bangor, work on the Conway Bridge had been proceeding rapidly and when all the Menai chains were in place, men and tackle were transferred to install the Conway chains. With a span of 327 feet between the suspension towers, Conway Bridge was a great work which would have received more notice from contemporaries had it not been overshadowed by the Menai Bridge. It was one of the rare examples (Tongueland and Craigellachie were others) in which Telford departed from that severely functional style which he adopted for nearly all his bridge work. In deference to the proximity of Conway Castle, Telford designed suspension towers in the form of castellated medieval gateways. Like Wolfe Barry's later

Tower Bridge where the presence of the Tower of London exerted a similar influence, the romanticism of Telford's Conway Bridge has since been the subject of harsh criticism from architects of the functionalist school. This is a very controversial question. There is truth in the functional argument but it is limited and like all limited truths it can easily be carried to excess. The evidence of this is painfully apparent in too much of the architecture and civil engineering work of today. Such uncompromising functionalism repels us by its arrogant contempt for its surroundings, and if it be said that Telford went to the opposite extreme of false compromise, at least he displayed an intelligent respect for the past which is sadly lacking today. Many would say that he did succeed in his intention of welding bridge and castle into one harmonious composition.

The heaviest work at Conway was not the building of the bridge itself but the formation of its approach embankment over the sands of the tidal river. Two thousand and fifteen feet long and 300 feet broad at the base, this was a work comparable in magnitude with the Stanley Sands embankment in Anglesey. On the site of the bridge the tides run so swiftly through the narrow channel that to use a raft for carrying the chains across as at the Menai would have been a difficult and dangerous undertaking. Fortunately, therefore, thanks to the shorter span, a different and less spectacular method of slinging them was evolved. Twelve of the $6\frac{1}{2}$-inch ropes used for raising the Menai chains were stretched across the river from tower to tower at the correct curvature and upon them a stout timber platform was built on which the chains could be assembled link by link.

Once all the chains had been successfully slung, the building of the suspended roadway platforms was a straightforward and comparatively rapid undertaking and both bridges were opened to traffic in 1826, the Menai at the end of January and the Conway on 1 July. It is characteristic of Telford, who hated pomp and circumstance, that the opening of the largest bridge in Britain was not attended by any ceremony or officially organised junketing. No major engineering work in history can have had a stranger and less ostentatious inauguration, for it took place in wild weather in the dead of a winter's night. For those few who took part in it, however, the drama of the occasion must have remained with them always.

It was after midnight on 30 January, a pitch dark night and blowing hard, when the Down Royal London and Holyhead Mail came over the Nant Ffrancon Pass. David Davies held the ribbons and the guard was William Read. Waiting for it on the outskirts of Bangor was William Provis and he stepped into the road when he heard above the soughing of the wind the cry of Read's post horn. Davies reined in his horses and in the glimmering light of the coach lamps Provis clambered aboard. The next stop was at the Bangor Ferry Inn. Here quite a party had been keeping vigil, determined to be amongst the first to cross the bridge by coach. There was Akers, the mail coach superintendent, William Hazledine, John Provis, Thomas Rhodes, the two young Wilsons and as many more as could find a precarious foothold on the coach. Round the bend of the road the lights of the bridge, special sperm oil lanterns made by James De Ville of London, starred the darkness over Anglesey and threw serpentine reflections on the storm-tossed water far below. So, at 1.35 a.m. on that winter morning while the great chains overhead stirred uneasily and the wind howled through the suspension rods, the first coach rumbled over the Menai and the bridge was opened.

No matter what Telford might think and notwithstanding the fact that daylight brought heavy rain, the local inhabitants were not going to allow the occasion to pass without celebration. Soon after it was daylight a procession of carriages and coaches began to cross the bridge. In the first carriage rode A. E. Fuller, one of the Commissioners, and in the second Telford accompanied by Sir Henry Parnell. Then followed three coaches, the Bangor stage coach *Pilot*, the Caernarvon day coach and the first London stage coach, the *Oxonian*. Next came Sir David Erskine driving himself in a carriage drawn by four splendid greys decorated with ribbons. He headed a seemingly endless procession of vehicles of every kind from the gleaming carriages of the local gentry to farmers' gigs and small pony carts. At mid-day the weather cleared and all that afternoon wheeled traffic and foot passengers moved over the bridge continuously while bands played, cannon crashed out and innumerable flags fluttered in the wind.

It would seem that, having crossed the bridge first thing in the morning, Telford drove straight on to Shrewsbury, for the next day William Provis addressed the following vivid account of the subsequent proceedings to him at the Talbot Hotel. 'The concourse of people who passed over yesterday was immense,' wrote Provis. 'At one time the Bridge was so crowded that it was difficult to move along. Most of the carriages of the neighbouring gentry, stage coaches, Post Chaises, gigs and horses, pressed repeatedly over and kept up a continuous procession for several hours. The demand for tickets was so great that they could not be issued fast enough and many in the madness of their joy threw their tickets away that they might have the pleasure of paying again. Not a single accident nor an unpleasant occurence took place and everyone appeared satisfied with the safety of the bridge and delighted that they could go home and say "I crossed the first day it was opened." The receipts were about £18.

'A good dinner having been provided at Mr. Jackson's, a party assembled there to make themselves merry and drink "success to the bridge". William Hazledine, Esq., having been called to the chair, kept up the life and spirit of the evening with his wit and funny stories; and there being a general feeling to receive all sorts of good things, they were soon as happy and joyous as could be wished.' Evidently William Provis shared this general feeling by consuming his fair share of the good things, for he concludes: 'As far as I can learn, all went off well, but it is difficult today to know what was going on yesterday.'

On the 2nd of July following it was young John Provis's turn to describe to his chief the opening of the Conway Bridge. He wrote: 'Conway bridge was opened yesterday morning between 12 and one o'clock by the Chester Mail with as many passengers as could possibly find a place about it that they could hold by. The horses went on steadily over which was more than I expected they would as the people were shouting and waving by the side of them from the embankment to the pier, the passengers at the same time singing "God Save the King" as loud as they could. . . . I could not hear that the slightest accident happened with the exception of a few broken windows at the public houses . . .' So, amid scenes of revelry and rejoicing which he did not stay to share, Telford's great roads to Holyhead were completed and Robin Ddu's prophecy fulfilled. A month later, Telford celebrated his seventieth birthday.

9

THE COLOSSUS OF ROADS

SOUTHEY's punning nickname is appropriate, not only for Telford the builder of roads and bridges but, in a different sense, for Telford the tireless traveller. 'Telford's is a happy life,' wrote Southey enviously, 'everywhere making roads, building bridges, forming canals and creating harbours – works of sure, solid, permanent utility; everywhere employing a great number of persons, selecting the most meritorious, and putting them forward in the world, in his own way.' Just as the small boy longs to be an engine driver, so all of us at some time or another envy the other man's job as something more romantic and less humdrum and beset by worries than our own. As in Southey's case, such envy is based on the superficial impressions of ignorance. Southey knew nothing whatever about engineering and there could have been few men then living, let alone Southey, who could for so many years have maintained so relentless a pace as did Telford.

And yet Southey was right; Telford's was a happy life because he was complete master of it; the stupendous amount of work which he performed was not a burden imposed upon him by others nor was it undertaken for any fanatical principle or ulterior motive such as a consuming ambition to achieve great wealth or worldly honours. He worked for the best of all reasons – because he enjoyed it. For this reason he could speak of his work as though it was some form of self-indulgence. Thus, when he had committed himself to building the Menai Bridge in 1818 he wrote to von Platen: 'I keep talking of avoiding employment, but when undertakings of magnitude and novelty are forced upon me the temptation is too great to resist.' Remember, too, that other saying of his to the same friend: 'There is a risk of the *Game* being over for us both. We *play* with so much eagerness – it cannot last long . . .'

A modern writer has described Telford as 'dedicated to the God of Work', a description which conjures up a false picture of a dour-faced

Scot whose life was one long masochistic frenzy of work and worry and who would sourly condemn any form of pleasure or self-indulgence as a sin. The truth is that today work has become such a mechanical drudgery, such a soulless and specialised routine for all but a very small minority that the notion of work as a pleasure and an end in itself has become almost inconceivable to the modern mind. If the wheels of our civilisation are running down it is for this reason.

Work for Telford was no obsession and consequently it never robbed him of his humanity; never consumed him. Few men of his generation could have boasted a wider or more varied circle of friends from Members of Parliament, Government officials, great landowners, men of science and fellow engineers to labourers, working craftsmen and country innkeepers. On his Highland journey with Telford, Southey repeatedly remarked on the warmth of the welcome they received at every stopping place. Everywhere Telford appeared to be well known and so well loved that his coming was an important event. With most of the innkeepers and 'locals' whom they encountered in the Highlands he appeared to be on Christian or nickname terms. He would chaff them, remember the appropriate stock jokes and afterwards regale Southey with local information, history and gossip.

It is clear from this that Telford took a consuming interest in the doings of his fellow-men and that his work, so far from inhibiting this, enabled him to indulge it by means of the constant travel which it involved. As he confessed to Andrew Little, travel suited his disposition and he would doubtless have agreed wholeheartedly with the maxim that it is better to travel hopefully than to arrive. It was this capacity of his to enjoy, year after year, a completely nomadic existance which is the secret of the prodigious amount of work which he managed to cram into his life. The preceding chapters have been concerned almost exclusively with a few of his most celebrated achievements and have given little indication of the great number of lesser undertakings which were claiming his attention at the same time. A mere list of them would cover a dozen closely printed pages and it is not the intention of this chapter to do more than to notice a few of the more significant before passing on to the last major works of his life.

Even in these days of rapid travel by land, sea and air it would be difficult for any one man to equal Telford's record in the personal supervision of engineering undertakings going on simultaneously in many remote parts of Britain which are still, even today, considered inaccessible. The following astonishing letter which he wrote to William Little from Glasgow in October 1816 when he was in his sixtieth year helps to explain how he achieved it.

'I have been and shall still, for some time, be much hurried,' he writes. 'After Parliament was prorogued, I went thro' North Wales where about 500 men are employed, & from thence into and along all the Eastern side of Ireland from Waterford to Belfast & Donaghadee and across by Portpatrick to Carlisle, from thence to Glasgow and back by Moffat to Edinburgh. From thence by St. Andrews & Dundee to Aberdeen, then up the Western parts of that County, then across it and Banffshire and to every Town on the Coast to Inverness – thence thro' Ross & Sutherland & back to Inverness – from thence across to Fort William on the West Sea and back to Inverness – then back to Fort William. From thence by Tyndrum & Inverary down to the Crinan Canal in Cantyre, thence back by Inverary and Loch Lomond to Glasgow and again still nearly the same route and back to Inverness and west by the Crinan Canal and Glasgow to perform over again before I reach the Border. This will give you some notion of my restless life, and, at this Season, a Post Chaise can scarcely render it Comfortable – the Weather here more than half Rainy and already much Snow on the Mountains . . .' Even when he had completed this itinerary the season's travelling was not over. In company with Alexander Easton he carried out a re-survey of the Glasgow and Carlisle road in heavy snow before returning at last to London at the end of November.

Records do not reveal the reason for his Irish visit on this occasion, but it was presumably to do with the improvement of the east coast harbours. He had submitted proposals for harbour improvement works at Bangor (County Down) and Donaghadee in 1808, while he later took charge of the works begun by Rennie at Howth and of the improvement of the road between Howth and Dublin. As long ago as 1800 he had written enthusiastically to Andrew Little about the prospect of his becoming engineer to the Board of Control for Ireland's inland waterways, but he did not get the post. Later he probably counted this a fortunate escape for the Board's record was a dismal one. His only connection with Ireland's inland waterways was in 1826 when he became consulting engineer to the Ulster Canal Company. Telford evidently did not like Ireland, for after a visit to Dublin in 1817 he commented tersely: 'I ... spent six weeks in the Metropolis of Pat's Country; I confess it enables me to set a higher Value on our own.'

Throughout the decade 1820–30 an immense amount of road survey work was done under Telford's direction, but unfortunately for his posterity the improvements he recommended were not carried out. The Government would not grant the money and with the advent of railways Telford's plans were forgotten. His assistants spent no less than eight years surveying the whole of the Great North Road from London to Edinburgh. In his autobiography, Telford summarised the recommendations which he based on these surveys in the following words: 'a mail coach road of the most perfect construction and unqualified straight line from London, by Barnet, Shefford, Newark, York, Newcastle, Morpeth, Wooler and Coldstream to Edinburgh may be reduced to 362¼ miles; whereas the present hilly and incommodious road is at least 391½ miles. But no authority less than that of Parliamentary Commissioners must be expected to accomplish the improvement on so large a scale.' The reconstruction of the last eleven miles of road immediately south of Edinburgh and the building of his new bridge over the Wansbeck at Morpeth in 1832 were the only works which Telford lived to see carried out and the inadequacy of Britain's premier highway has remained a monument to the shortsighted parsimony of successive Governments from that day to this.

In 1823 a service of steam packets was about to be introduced between Milford Haven and Waterford. As a result the Postmaster-General instructed Telford to survey the mail roads from Northleach and from Bristol to Carmarthen, where they united, and from thence to Milford. West of Northleach the existing route passed through Cheltenham, Gloucester, Ross, Monmouth and Abergavenny to Brecon, but there was considerable local feeling in favour of an alternative and the people of Cheltenham went so far as to petition the Chancellor of the Exchequer in these terms: 'The Town of Cheltenham has of late years greatly encreased in Magnitude and Extent and is still increasing and as appears by the last census contains a population of 13,396, greatly exceeding that of the City of Gloucester. That the Town is very much resorted to by numbers of Irish Families of the First Respectability many of whom come from the Southern parts of that Division of the British Empire and that in consequence thereof it is a matter of great and vital importance to Cheltenham that the communication between that town and the South of Ireland should be rendered as short and commodious as is practicable.' The petition went on to explain that an Act for the new Haw Bridge over the Severn would be presented to the present parliamentary session and that this bridge, with the new connecting roads which they proposed to build from Piff's Elm, near Combe Hill, westwards would reduce the distance to Hereford to thirty-four miles. The petition concluded: 'that the road through Hereford, the Hay and Brecon is nearer, more level and better adapted for safe and speedy communication with Milford Haven'.

Having examined both routes, however, Telford pronounced that the difference in distance was only $3\frac{3}{4}$ miles and he finally decided against the Cheltenham scheme. Oddly enough, notwithstanding this adverse verdict, Haw Bridge was sanctioned and built, but not the proposed new road connections. Consequently it remains to this day almost the only bridge across a major river in England which carries only local, rural traffic because, lost in a maze of by-roads it has no other apparent purpose. Many have speculated upon the motive of its builders, not realising that it is a monument to Cheltenham's long-forgotten dream of a new Anglo–Irish route.

The Bristol–Milford mail route crossed the Severn estuary by the New Passage Ferry and then passed through Newport, Cardiff and Swansea. In his report, Telford described New Passage as 'One of the most forbidding places at which an important ferry was ever established – a succession of violent cataracts formed in a rocky channel exposed to the rapid rush of a tide which has scarcely an equal upon any other coast'. He recommended twenty-two miles of new road from Bristol to Uphill Bay, a ferry service to Sully Island and another new road from thence to join the existing road from Cardiff to Cowbridge, but the Postmaster-General would not accept this proposal. Another suggestion of Telford's was a suspension bridge over the Severn at Beachley, but the only improvement of the river crossing which was actually carried out at this time was that of the Old Passage Ferry between Beachley and Aust. Here new ferry piers were built by Henry Habberley Price, Telford acting as consulting engineer. Steam boats were introduced at both New and Old Passage ferries and the London–Chepstow mail coaches began to use the Old Passage from 27 August 1827 onwards.

The South Wales road survey was carried out by Henry Welch and Joseph Mitchell, who was temporarily transferred from the Highlands. Telford's instructions to Welch survive to show what he expected from his assistants on work of this kind. He was to prepare five-inch-to-the-mile maps of the existing road and the proposed improvements. Also sections to a horizontal scale of ten inches to the mile and a vertical scale of 100 feet to the inch. He was to indicate the width of the road on the map, give occasional cross-sections, mark all bridges and cross drains and their dimensions, describe the condition of the surface, the materials used, where obtained and at what expense. Where sections of new road were proposed he was to note the availability and expense of road material, the dimensions of any bridges required, the landowners' and occupiers' names, the county, parish and township. The work was to be completed in six months and Welch was to be paid at the rate of three guineas a mile plus a guinea per day expenses.

A letter from Welch to Telford written in August 1824, when he was surveying a proposed deviation from the existing route between Lea and Ross, shows that he had to contend with as much local opposition as did the railway surveyors some years later. 'I commenced to survey,' he writes, 'but was stopt after receiving much abusive language from going thro' the corn. I then went upon the Abergavenny Rd until I thought the fields would be cleared, then returned and resumed the variations. . . . I determined not to be easily removed a second time. One field of wheat, however, remained only partially cut when I reached it. My men were obliged to leave the field otherwise they would have been ill treated by the reapers who were mostly intoxicated and Lawless Fellows from the Forest. An instance of their effrontery was some of them lifted my Portmanteau containing the letters you sent me and threw it into a Field of Barley where it remained, supposed to have been stolen, until the Barley was cut.'

Poor Welch, he discovered to his cost that the men of Dean Forest were a law unto themselves and it is small wonder that his survey took him, not six months, but two years. Moreover, he laboured in vain for, like the Great North Road, the defects of the road to Milford were presently forgotten amid the excitements of the railway age.

Though Telford was disappointed in so many of his English and

Welsh road-building schemes, he lived to build many more bridges south of the Border. Another of Southey's nicknames for Telford was 'Pontifex Maximus', and it is certainly true that he was the most prolific builder of bridges this country has ever known. In addition to the prodigious number of Scottish bridges and excluding those constructed in connection with his canal works, Telford built a total of twenty-seven important road bridges in England and Wales. Among these were three more bridges over the Severn: the Mythe Bridge at Tewkesbury (1823–6), Over Bridge across the Maisemore Channel of the river at Gloucester (1825) and Holt Fleet, near Ombersley (1827).

There was no previous bridge over the Severn at Tewkesbury. An unnamed architect[1] prepared a design for a bridge of three iron arches and an Act for its construction was obtained in 1820, but the promoters subsequently fell out with their architect and in 1823 Telford was called in. He designed an iron bridge with a single span of 170 feet approached on the west side by an embankment and an extended masonry abutment pierced by a series of narrow arches to allow the passage of flood water. Hugh McIntrop was the contractor for the embankment and abutments, while the arch was of Shropshire iron, the work of William Hazledine. It was opened in April 1826. Why it took so long to build is not clear, as no particular difficulties have been recorded. Perhaps the promoters ran short of money. Complete with its charming miniature toll house at the eastern approach, the Mythe Bridge survives intact as a splendid example of Telford's iron bridge work.

Over Bridge, which replaced a medieval bridge on the same site, is notable in two respects. First, its design was most unusual and secondly it is the only example of Telford's bridge work in which, as he freely admitted, he was guilty of a serious error.[2] Telford first designed for Over a single-arch iron bridge similar to that at the Mythe except that the span was reduced to 150 feet. The Gloucester magistrates, however, insisted that they wanted a bridge of stone, perhaps because the nearby bridge over the Gloucester channel of the Severn (Westgate Bridge)

---

1. It may be that he was the architect of Haw Bridge and that his Mythe design was adapted for use there. Certainly Haw Bridge is a clumsy piece of work, but this is pure speculation.

2. Settlement of the piers of the Don bridge at Aberdeen, built by John Gibb (1827–30) gave Telford a great deal of trouble but appears to have been due to circumstances which could not have been foreseen.

had been replaced in stone by Sir Robert Smirke[1] between 1814 and 1816. Telford then designed a single arch of stone based upon that which the French architect Perronet had evolved in 1768 for his bridge over the Seine at Neuilly. The body of the arch was an ellipse with a chord line of 150 feet and a rise of thirty-five feet, but the voussoirs, or external arch stones, were set to segments on the same chord with a rise of only thirteen feet. The effect of this is that instead of being four-square, the appearance of the arch is funnel-shaped, this shape being most pronounced at the springing and tapering off to the crown. Thirty years' acquaintance had left Telford with no illusions about the power of the swollen Severn and the object of his unusual design was that such an arch would offer less resistance to flood water.

Contractor for the bridge was John Cargill, erstwhile partner of the late John Simpson on the Caledonian Canal, Bonar Bridge and other works in the Highlands. The foundations were formed in the same fashion as those at the Mythe. The riverside meadows were excavated until, at depths of twenth-seven feet on the west bank and thirty-three feet on the east, a bed of natural gravel was encountered. Rubble stone was laid on this gravel and then great baulks ('sleepers', Telford calls them) of Memel pine between thirty-seven and forty feet long topped with beech planking. Upon this timber platform the first stone courses were laid, some of the blocks weighing as much as three tons. The stone was brought down the river from the quarries at Arley where, so many years before, Telford had found the stone for his Bewdley Bridge. Unfortunately, however, the same care was not taken in making the foundations for the abutment wing walls. The result was that when the centering of the arch was eased the wing walls yielded to the thrust and the crown of the arch dropped ten inches. 'I much regret it as I have never had occasion to state anything of the sort in any other of the numerous bridges described in this volume,' wrote Telford in his autobiography. 'I more especially take blame to myself for having suffered an ill-judged parsimony to prevail in the foundations of the wing-walls, leaving them unsupported by piles and platforms, because if so secured, I am convinced that the sinking of the arch would not

1. Smirke was a highly successful architect whose work in the neighbourhood included Gloucester Shire Hall and Eastnor Castle. He was not so successful as a bridge builder for his only other known bridge, that over the Eden at Carlisle, collapsed.

have exceeded three inches.' Notwithstanding this blow to its designer's professional pride, however, Over Bridge stands to this day beneath a weight of traffic undreamed of by Telford. When it was opened, its unusual construction was highly praised as 'giving a character of airiness and lightness to the arch', but latterly it has been the subject of adverse criticism with which this writer does not agree. In his Wansbeck Bridge at Morpeth Telford used the same unusual design of arch.

In one of the two last and greatest of the stone bridges designed by Telford, the Dean Bridge over the Water of Leith at Edinburgh, the arches are once again of unusual design. Foundation difficulties forced Telford to modify his original three-span design for the Dean Bridge to one of four arches, each of 90 feet span and with a maximum height above water of 106 feet. The main arches carrying the roadway rise 30 feet, but the footpaths upon either side are supported on secondary arches which have a span of 96 feet, a rise of only 10 feet, and spring from the piers at a height 20 feet above the springing of the main arches. It seems likely that it was the praise which greeted his Over Bridge that led Telford to adopt this design, for although the effect is similar it has no practical motive. On the other hand, by masking the more massive main arches, the delicate footway arches do impart a deceptive impression of lightness to the structure. In order to reduce weight, the tall piers are one of the finest examples of that hollow wall construction which Telford first used at Pont Cysyllte. Charles Atherton was Telford's resident engineer for the Dean Bridge and John Gibb of Aberdeen the contractor. It was completed in December 1831. After so great a number of suicides had flung themselves from it that it was nicknamed 'the Bridge of Sighs', the parapet was heightened.

Pathhead Bridge, carrying the Great North Road over the Tyne Water at Midlothian, was a smaller version of the Dean Bridge and was completed in the same year. It is 68 feet high and has five spans of 48 feet.

The same team were responsible for building the Broomielaw Bridge over the Clyde at Glasgow, the second of the two bridges alluded to and Telford's last important bridge work. It was not until November 1832 that he signed the contract drawings and although from the commencement in March 1833 the work was carried on with remarkable speed, he did not live to see its completion on New Year's Day, 1836. It is sad that Telford's Broomielaw Bridge had to be demolished when the Clyde was deepened in the 'nineties, for illustrations reveal that it must have been the most beautiful of all his stone bridge works. It was a beauty of symmetry and proportion for, unlike the Dean Bridge, the design was completely orthodox. There were six spans of 52 feet each and a centre span increased to 58 feet 6 inches, which was just sufficient to give it the necessary emphasis. It complied with Telford's dictum that the proper elevation of a bridge should be a segment of a circle springing from the abutments, although here the circle was a vast one, the total rise to the centre (exaggerated in contemporary illustrations) being only 2 feet 9 inches. Shallow buttresses framed each arch and the elegant stone balustrade was perfectly proportioned.

Telford did not live to build any more suspension bridges. Structural weaknesses which soon became apparent in his great bridge over the Menai had the effect of making the old engineer cautious and disinclined to venture upon new and greater flights. Only a week after it was opened, on the night of 7 February 1826, a terrific gale swept through the Straits and, striking the bridge broadside, set up movement so violent that the coachman on the Down *Oxonian* refused to cross. 'The bridge certainly laboured very hard', Thomas Rhodes admitted in his report to Telford, 'and the night being dark and the wind whistling thro' the railing and chains made it appear rather terrific.' He later reported twenty-four of the roadway bars and six suspension rods broken.

On 19 February an even more violent gale blew up in the night and John Wilson reported twenty more suspension rods broken and another fifty bent. Everyone admitted that the undulating movement of the bridge platform in a high wind was very alarming. On this occasion the Chester mail waited at the bridge approach for an hour and a half in the hope that the storm might abate, but it eventually crossed safely. On 1 March, however, the same coach was less fortunate. Guard Lumley's time bill reads: 'Twenty minutes lost at the Bridge it blowing hard and the Bridge in great motion which caused the horses to fall all down together and be intangled in each other's harness. Had to set them at liberty one at a time by cutting the harness.'

Obviously such a state of affairs could not continue and various modifications were made. Stronger suspension rods were fitted and the method of attaching them to the roadway bars was altered. Bracing chains were introduced to check the movement of the main chains and, later, the timber platform itself was strengthened. That troubles of this kind should develop in a structure of such unprecedented size and novelty was really not surprising and they were not serious enough to impugn the basic soundness of Telford's design. They represented a challenge which a younger man would have been eager to accept and to conquer, but their effect upon the ageing Telford was quite otherwise.

Two years later, the committee appointed to build a bridge over the Avon Gorge at Clifton, near Bristol, invited Telford to judge their competition for a suitable design. Confronted by the drawings for a suspension bridge with a span of 916 feet which had been submitted by a young man named I. K. Brunel he shook his head doubtfully. The fact that he had once designed a bridge of even greater span was forgotten. As a result of his experience at the Menai he had become obsessed with, the problem of resisting lateral wind pressure and the possibility that this young man, thanks to his father's tutelage, might have produced an answer to it did not seem to occur to him. He expressed the opinion that his Menai Bridge represented the maximum safe span for a suspension bridge and dismissed Brunel's design along with those of all the other contestants.

At the request of the Bridge Committee, Telford then produced a design of his own in which he proposed to reduce the length of the suspended span by means of two great piers, decorated in the Gothic style, rising from the floor of the Avon Gorge. This, he explained, 'guarded against the effects of violent gusts of Winds which must be expected in that deep and narrow valley'. This design was at first greeted with great acclamation, but young Brunel, undaunted by Telford's age and eminence, made no secret of his contempt. 'The idea', he wrote, 'of going to the bottom of such a valley for the purpose of raising two intermediate supporters hardly occurred to me,' and he went on to express the opinion that such timidity cast a poor reflection on the state of the Arts. The effect of this withering criticism was that enthusiasm for Telford's proposal quickly evaporated. The Bridge Committee decided to hold a second competition in which Telford was decisively defeated and Brunel triumphed.

We may be sure that if, from some corner of Elysium reserved for engineers, Telford looks back upon the varied achievements of his long life, he would wish to consign to merciful oblivion both his Runcorn and Clifton designs. The first was ill-considered and over-bold; the second a product of the faint-hearted conservatism of old age. Upon the other hand, if there is one work by which he would wish to be remembered by posterity it is surely that great design for a new London Bridge which he first put forward in 1800 when he was in his self-confident prime and sure of his mastery of iron. Like the Runcorn suspension bridge, this London Bridge was conceived upon a breathtaking scale, but it is the *tour de force* of a man at the height of his power and it has that quality of greatness which the Runcorn design altogether lacks.

It is impossible to look at this proposed London Bridge without bitterly regretting that 'Merlin' Hazledine and his attendant wizards from the Welsh Border were never given the opportunity to display their powers on Thames side. For Telford planned to bridge London river with a single iron arch of 600 feet span having a height above high water at the crown of 65 feet. The roadway was to be 45 feet wide and the weight of the iron arch 6,500 tons. It was indeed a *Great Eastern* among bridges, and the strange thing is that it does not seem to have excited the ridicule and incredulity which usually greeted proposals so unprecedented and which, as we have seen, the Menai Bridge design duly aroused. There was certainly considerable controversy, but the consensus of opinion among engineers and men of science so eminent as Watt, Jessop, Wilkinson, Reynolds and Professors Playfair and Robison was that the bridge was a practicable proposition. According to Smiles, preliminary works on the riverside were actually begun, but it was not to be. Apart from the unfavourable economic climate of the time, the promoters soon ran into the difficulty which has bedevilled every attempt to bring grace and space to the City of London from the time of the Great Fire to the present day. In order to secure an ample headroom for shipping without an inordinate rise to the crown of the arch, Telford's design called for high-level approach ramps of masonry in the form of colonnades and, as his engraving shows, these would have extended along the waterfront. George Dance the younger, architect of Newgate Prison and at this time Clerk of City Works, was to have executed these approaches, but they were the downfall of the scheme on the score of the prohibitive value of the ground they would occupy. So Telford's bridge remained a dream although even on paper it enhanced his reputation. Copies of his design were printed for sale to the public and it is evident in the bills he received from his bookseller, Joseph Taylor of Holborn, which show his print sales *per contra*, that they were still selling briskly ten years after publication. Were it not dated by the shipping in the river, Telford's picture of his mighty bridge would still appear, even today, as some apocalyptic vision of the future.

The replacement of old London Bridge was only a part of a comprehensive plan for the development of the Port of London which George Dance had drawn up, and Telford was frequently consulted by the Select Committee appointed to investigate these proposals. Dance's scheme was never carried through, however, and in the building of London's first wet docks Telford played no part. William Jessop was responsible for the West India Docks which were begun in 1799, and John Rennie for the London Docks, begun in 1802 and completed in January 1805. Rennie was also responsible for the building of the East India Docks in association with Ralph Walker. It was Rennie too, who, just before his death, designed the new London Bridge which was built by his sons. Yet Telford lived to make his mark on London's dockland.

The new London Dock Companies soon became extremely powerful vested interests thanks to rapidly expanding commerce, and by 1820 their exorbitant charges had become notorious. The Warehousing Act of 1823 which allowed goods to be stored in bond, duty free, brought to a head the movement in favour of the building of a new 'free' dock which would break the vicious monopoly of the old Companies and provide badly needed new accommodation in the overcrowded port. The site of the ancient St. Katharine's Hospital near the Tower of London was selected as being conveniently close to the City and in 1824 the St. Katharine's Dock Company was formed. The Bill of Incorporation met with violent opposition from the old companies, but in face of the new dominant philosophy of free competition this proved unavailing and the new Dock Company obtained its Act in June 1825. The project involved the demolition of no less than 1,250 houses in addition to the old hospital, and, since the total area of the site was only twenty-seven acres, we can picture what a crowded rookery it must have been. Telford was appointed engineer to the Company and the site was cleared ready for work to begin by May 1826. This date fitted in well with Telford's arrangements, for it will be remembered that the Menai Bridge had been completed in January of that year so that out of the trusty band of assistants who had been working on the bridge he was able to pick Thomas Rhodes as his resident engineer at St. Katharine's.

Telford was faced with the difficult problem of fitting the maximum amount of accommodation into a very restricted and awkwardly shaped site, indeed when space had been allowed for warehouses the area available for the dock basins was only ten acres. In adapting these basins, two in number, to the site, he therefore abandoned the rigid rectangular shapes adopted by Jessop and Rennie in forming the earlier docks. He designed an entrance basin and a lock from the river 180 feet in length and 45 feet wide. This lock was provided with the usual sluices or 'paddles', but owing to the small area of the dock some means had to be provided for maintaining the level of the basins and compensating the loss of water consumed in lockage. Two 80 h.p. Boulton & Watt beam engines driving six double-acting pumps were therefore installed. They drew water through a culvert 170 feet long which extended to the centre of the river and could deliver it either into the basin, into the lock or into both simultaneously. Using sluices and pumps together, the entrance lock could be filled in five and a half minutes. The reason why Telford designed two docks instead of one despite the small area, was that one dock at a time could be closed for cleaning or 'scouring' without closing the dock, since the entrance basin enabled either to be used at will. He set the lock sills as low as possible in order to prolong the period of entry on each tide.

Construction of the dock proceeded with such extraordinary speed that the whole of the work, basins, warehouses, entrance lock and pumping plant, was completed on 25 October 1828, only two years and six months from the laying of the foundation stone. Such immoderate haste disquieted the ageing engineer. 'As a practical engineer,' wrote Telford in his autobiography, '... I must be allowed to protest against such haste, pregnant as it was, and ever will be, with risks, which in more instances than one severely tasked all my experience and skill, involving dangerously the reputation of the directors and their engineer.' All his long life Telford had been accustomed to take the utmost precautions to safeguard the lives of his workmen, for his experience as a working mason had made him well aware of the risk involved in any large-scale building enterprise. The fact that none of his great works were marred by any serious accident involving loss of life

shows how well he succeeded. There does not appear to have been any
disaster during the building of St. Katharine's but, at seventy-one,
Telford was now finding himself in an age where commercial oppor-
tunism overrode such scruples and in which there was growing up a
new generation of engineers who lacked his background of practical
experience, men accustomed to take risks and to hold the lives of their
workmen of small account. The railway age, in fact, was near at hand.
The Dock Company's architect, who collaborated with Telford in the
design of the warehouses, was Philip Hardwick, a man thirty-five years
his junior who would presently achieve fame as the designer of Euston's
great Doric portico.

In his old age Telford spoke continually to von Platen and others of
the need to cut down his commitments. But apart from his own self-
confessed predilection for work, two circumstances combined to
prevent him from doing so. One was his appointment as consulting
engineer to the Exchequer Loan Commissioners in 1817 and the other
was the death of John Rennie in 1821.

A financial slump followed the peace of Waterloo, in which it
became almost impossible to raise capital for new ventures. Consequently a great many projects came to a standstill for lack of means
and there was a growing risk of serious unemployment. It was to counter this that the Government set up the Exchequer Loan Commission
in 1817 at the instigation of Vansittart, the Chancellor of the Exchequer,
and empowered it to make capital loans to deserving undertakings.
The Commission continued in being until shortly before Telford's
death and throughout these years it was his duty as the Commission's
engineering adviser to weigh the merit of any loan application received.
As a result there was scarcely an engineering undertaking of any
moment in the country which did not at some time or other claim his
attention.

In considering the formidable list of projects to which loans were
granted it is not always easy to determine the degree of Telford's par-
ticipation. In many cases it would seem that he was content to approve
a scheme on paper and these need not concern us. In others he would
send one of his assistants down to examine and report upon the state of
the works before making his decision. Thus Alexander Easton was sent
to Cornwall in 1823 to investigate the Bude Canal with its inclined plane
lifts. His confidential reports to Telford disclosed a curious state of

affairs not altogether to the credit of James Green, the Company's engineer. Nevertheless, the Bude Canal was granted a loan of £20,000.

There were yet other cases where an undertaking, though deserving financial aid, had got itself into difficulties so serious that Telford intervened personally to such an extent that he may fairly be awarded at least a share of the credit for the finished work. Of this group by far the most important was the Gloucester & Berkeley Ship Canal.

The Gloucester & Berkeley was a project that had been in difficulties ever since it was conceived in the canal mania period. It was an ambitious scheme designed to by-pass the dangerous estuary of the Severn. An admirable detailed account of its long and chequered history will be found in Mr. Charles Hadfield's *Canals of Southern England*. It is sufficient here to say that of the many engineers who at one time or other had a finger in this unprofitable pie, Robert Mylne was the man chiefly responsible. There are some striking similarities between Robert Mylne's career and that of Telford. Mylne came of a family of Scottish stonemasons and he began his career as a working mason in Edinburgh. Like Telford he aspired to become an architect and turned to civil engineering. Unlike Telford, however, he endeavoured to bestride both professions, combining his engineering exploits with a considerable architectural practice. The former seem to have suffered in the process, with the consequence that it is as an architect that he is remembered and his only notable work in Telford's field was his Blackfriars Bridge. For cutting a canal 17¾ miles long, 70 feet wide at top and 15 feet deep from Gloucester to Berkeley Pill, Mylne had estimated that £121,329 10s. 4d.[1] would be required, a sum notable for its utter inadequacy even in that most optimistic period of the canal boom. Needless to say, the Company were soon in serious financial difficulties, while the working committee complained bitterly of Mylne's inattention, declaring that he did little to earn his salary of £350 a year and should be paid by the day's work in future. To this Mylne was forced to agree, but early in 1798 he severed his connection with the Company. He died in 1811 at the age of seventy-seven.

When the Company appealed for a loan from the Commissioners the basin at Gloucester had been completed and in use for some years, but canal construction had proceeded no further than Hardwicke,

---

1. The suspicion that the shillings and pence were thrown in to give a false impression of accuracy is difficult to resist!

about a quarter of the total distance to Berkeley Pill. John Upton was in charge of the works. He had joined the Company as a clerk in 1811, but in the absence of anyone else he had assumed the office of engineer. Judging from the deeply aggrieved tone of a letter from Upton which has survived among Telford's papers, he had made himself decidedly unpopular. When Telford appeared on the scene John Woodhouse had just been appointed resident engineer, but he was soon succeeded by a Mr. Fletcher. Most probably this was the same J. Fletcher who was engineer of the Chester Canal and who subsequently partnered John Simpson in the construction of the Ellesmere Canal Company's extension from Whitchurch to Hurleston Junction. In that case there can be little doubt that he was appointed to the Gloucester & Berkeley on Telford's recommendation. 'You had condescended to meet me at the Bell,' wrote Upton to Telford from Gloucester, 'I was there to the minute – you was gone & on my return I saw Fletcher who wildly said you was gone and from his extraordinary demeanour towards me . . . I have conceived the idea that he wishes to prevent my seeing you – he has repeatedly said no one should ever be employed again – but why, what have I done to him or the Compy? If my abilities had not been more than equal to his, why is he afraid?' The unfortunate Upton evidently realised that so far as the Gloucester & Berkeley was concerned he had shot his bolt, for he concludes his letter by asking Telford whether he can find him a job on the South Wales road survey. He had apparently been guilty of selling building materials to himself as engineer. Racketeering of this kind seems to have been rife on the Gloucester & Berkeley, for John Woodhouse's dismissal was due to Telford's discovery that he had been buying on behalf of the Company building stone of inferior quality from his son. 'I am of the opinion', Telford informed the Company, 'that it is *absolutely necessary* to employ as Resident Engineer a person wholly unconnected with Contractors for Material or Labour in any shape.'

Having, as we would say, cleaned up the affairs of the Company, work on the canal went forward under Telford's direction with Fletcher in charge, and with the aid of very substantial loans from the Commission totalling no less than £160,000. The site of the southern junction with the Severn was changed from Berkeley Pill to Sharpness Point in accordance with a recommendation previously made by an engineer named Benjamin Bevan, and the depth of the canal was increased to eighteen feet. The extent of Telford's contribution to the Gloucester & Berkeley is difficult to determine, but that it was considerable may be judged by the fact that, alone of all the Loan Commission works which claimed his attention, he refers to the canal in a letter to von Platen dated 7 July 1823. 'The completing the Gloster Canal', he writes, 'is now also arranged, which is to cost £120,000. I have nearly completed the working drawings.' Unfortunately these drawings do not appear to have survived, nor do they feature in the Telford *Atlas* so that it is impossible to tell, for example, whether Mylne or Telford was responsible for the charming little bridgemen's houses which are such a feature of this canal. Mylne, most probably, but on the other hand it is almost certain that Telford designed the old canal entrance lock[1] at Sharpness Point with its attendant lockhouse. The Gloucester & Berkeley was finally opened in April 1827 and proved very successful. Unlike too many canals, it carries a good traffic to this day. It had cost altogether £444,000 and was the last ship canal to be built in England until the coming of the great Manchester Canal sixty-seven years later. Telford was consulted about two other projects of like kind, the Liverpool Ship Canal across the Wirral Peninsula and the English & Bristol Channel Ship Canal which was the last serious attempt to realise the old dream of a canal to obviate the long passage round Land's End. Of these the first proved to be a financial racket from which Telford retired in disgust, while the second was killed by steam power at sea and on rails.

When John Rennie died on 4 October 1821, Telford became the undisputed head of the civil engineering profession in Britain, and although Rennie's sons succeeded him and carried on many of his works in progress, including his London Bridge, the demand for Telford's services inevitably increased. Telford referred to Rennie's death and to the heavier burden which it had occasioned in a letter written to von Platen at the end of March 1822. This contains his first intimation that the pace was beginning to tell upon him. 'These great works,' he wrote, 'being scattered over all England, Scotland and Ireland, require much more travelling than is consistent with comfort and, if I am not careful, with health.'

Although they were fellow-countrymen and although in later life

1. New docks at Sharpness and a new entrance lock were opened in November 1874.

we find their names occasionally coupled, there is little evidence of any close personal collaboration between the two most eminent engineers of their day and they were certainly not friends. A curious letter written by Telford from Chester in April 1805 to James Watt at his home at Heathfield helps to explain the reason for this. Watt had evidently sent Telford the news of the death of Rennie's wife and Telford replies:

'I am truly sorry to find that Mr. Rennie has suffered so serious and distressing a loss and I am only sorry to inform you that his conduct prevents me from benefiting by his acquaintance. Altho' I never had any connection with him in business or ever intentionally did anything to injure or interfere with him, I, in every quarter, hear of his treating my character with a degree of illiberality not very becoming. This is so marked a part of his conduct that I really believe it does him a serious injury and proves serviceable to me. As I am desirous of not suffering in your good opinion, I mention this with a view to counteracting any insinuations which may be advanced to my disadvantage.'

From this it is obvious that there was no love lost between Rennie and Telford, although in the absence of any evidence other than this one letter it is impossible to judge the rights and wrongs of a matter of this kind. Only a contemporary such as James Watt, who knew both men well, could have done so. Although Rennie was Telford's junior by four years, he had risen to fame as an engineer earlier in life and perhaps at this time, when Telford was seriously threatening his preeminence, his professional jealousy was roused. In particular, he may have resented Telford's appointment as engineer to the Commissioners for Highland Roads and Bridges. True, he disdained road making but he may have felt that he had a better claim as a bridge builder. Certainly the beautiful classic proportions of Rennie's stone bridges display an architectural refinement lacking in Telford's more austere and severely practical designs, Bewdley and Broomielaw bridges alone excepted. On the other hand, in the use of cast iron no work of Rennie's can be compared with Bonar, Craigellachie or the Mythe. As his few iron bridges show – his bridge over the Witham at Boston for example – Rennie's handling of the new material was conservative. He did not, like Telford, adopt an entirely fresh approach, extracting the utmost from the new medium by refining away all unnecessary weight to achieve an effect so light and seemingly unsubstantial as to dazzle contemporary eyes. By contrast, Rennie's works look ponderous, being essentially stone bridges rendered in iron. As a canal engineer, Rennie carried out some impressive works but, despite Smiles's remarks to the contrary, his estimates were apt to be even more optimistic than Telford's and the subsequent history of his canals was singularly unfortunate. This may not have been the fault of the engineer except in one respect – almost without exception they suffered from inadequate water supply. Nevertheless, Rennie's deserved eminence was such that he had no good reason to grudge his countryman a place beside him.

The last of the few works with which the names of Telford and Rennie were both associated was carried out in that region which has at one time or another attracted the attention of practically every civil engineer of note – the Great Level of the Fens. This was the Eau Brink Cut which Rennie had planned for the purpose of cutting off a bend in the channel of the Great Ouse just above King's Lynn and so improving the river's outfall. This, like so many projects for the improvement of Fen drainage, aroused a prolonged and violent controversy which was finally settled by arbitration in 1818, Rennie representing the Commissioners for Drainage, Telford the Commissioners for Navigation and Captain Huddart acting as umpire. Agreement having been reached the work was put in hand, Telford still assisting actively, and was completed just before Rennie's death. It did not fulfil its intended purpose. Telford expressed the opinion that the dimensions of the channel decided by the umpire were too small and should be enlarged by a third. Rennie's son (later Sir John Rennie) agreed, and between them they succeeded in overruling Huddart. After this enlargement the cut was completely successful. Within one year the river bed between Denver and Lynn had been scoured to a depth of five feet, while the old loop of the river rapidly silted up and became pasture land.

Telford also helped Sir John Rennie to complete another of his father's plans, the cutting of the river Nene outfall from Gunthorpe below Wisbech to Crab's Hole Sluice on the Wash. This again, had been the subject of interminable argument. Rennie had produced detailed plans in 1814 but it was not until June 1827 that an Act was obtained. Telford described the administration of the Great Bedford Level as an incongruous establishment riddled with jealousy, contention and confusion, while on the subject of this scheme he wrote: 'There cannot be a stronger instance of inconvenience and delay experienced in bringing to bear any scheme of magnitude in which numerous and conflicting interests are to be consulted than in the case of the Nene Outfall.' But once the Act had been obtained the contractors, Jolliffe & Banks who had previously carried out the Eau Brink contract, went to work with a will, and the outfall was finished in 1830. One thousand one hundred men were employed and on Telford's recommendation William Swansborough of Wisbech acted as resident engineer.

For Telford's last work in the Fens he was alone responsible. This was the draining of the North Level, a great area of Fenland between Crowland and Guyhirn centring about Thorney. This he achieved by substituting for the old, narrow and winding Shire Drain a new, straight navigable cut, thirty-six feet wide, known as the New North Level Drain which ran from Clow's Cross to Gunthorpe Sluice at the head of the Nene Outfall channel. Begun in 1830, it was completed in 1834 and its effect was immediate and dramatic. Tycho Wing, the Duke of Bedford's agent, described how at Thorney one Sunday morning the whole congregation, parson included, trooped excitedly out of the church in the middle of the service on hearing the news that the long stagnant dykes had suddenly begun to flow.

It is not easy to envisage Telford in the flat landscape of the Fens. He was so essentially and in spirit a man of mountains that the Great Level must have seemed far more alien to him than the lakes and granite rocks of Sweden. Perhaps it was more than a coincidence that it was in the Fens that his health failed him for the first time. He was inspecting the Nene Outfall works in company with Sir John Rennie and John Gibb one day in July 1827 when the party was soaked to the skin by a heavy rainstorm. When they returned to their inn at Wisbech young Rennie retired at once to his bed and John Gibb prudently changed his clothes, but Telford was content to dry himself out by the fireside in the parlour as he had done, no doubt, many a time before after hard days spent in the mists and snows of the Highland glens or on the rainswept seaboard of west Wales. But this time, at the age of seventy, he tried even his iron constitution too far. He contracted a severe chill. On his road home to London complications developed and he became so seriously ill that he had to break his journey at Cambridge. Here he suffered the first bout of an internal trouble accompanied by violent bilious attacks which would recur and would ultimately prove mortal. Although he apparently made a rapid recovery from this first attack, John Rickman, who had known him well for so many years, maintained that he was never quite the same man again and lacked much of his old untiring energy. Nevertheless, Telford had still seven more years of life ahead of him during which he was able to see all but two of his last important works completed. Of these one was his Broomielaw Bridge. The other was his Birmingham & Liverpool Junction Canal which had been begun with every promise of early success in 1825 but which was destined to shadow his last years with anxiety, disappointment and bitter mortification.

THE COMING OF RAILWAYS

EVEN if Telford had not been preoccupied with the Caledonian and Gotha Canals, England would have afforded him little scope to display his powers as a canal builder during the first quarter of the new century. The many companies which had been floated with such enthusiasm and high hopes in the boom period of the early 'nineties fought gamely and with varying success to complete their canals in the face of steadily rising costs and shareholders who either would not or could not respond to calls. Meanwhile the old-established companies such as the Bridgewater or the Trent & Mersey sat tight in their enjoyment of a comfortable and highly lucrative transport monopoly. Their traffic was more than Brindley's narrow, tortuous waterways could pass efficiently. The combination of high toll charges with narrow, shoaled channels and towpaths where boat horses often struggled fetlock deep in mud stung canal traders to bitter protest, but it was a case of take it or leave it. Practically no improvement work was done. The only thing guaranteed to rouse the most prosperous and lethargic of canal proprietors to action was a competitive threat, however feeble, from some less prosperous and therefore less complacent canal company. The action of the Trent & Mersey in preventing the Chester Canal Company from building their branch to Middlewich and thus uniting the two canals was typical of a policy which effectually prevented the growth of an efficient national water transport system.

In these circumstances it is scarcely surprising that between 1795 and 1825 Parliament was called upon to sanction only three new English canal promotions of lasting importance, and with two of these, the Grand Union and the Regent's Canals,[1] Telford's name was indirectly

1. The Grand Union Canal connected the Leicestershire & Northamptonshire Union Canal at Foxton with the Grand Junction Canal at Long Buckby and so provided a through route between London, Leicester and the Trent. Although surveyed by Barnes and engineered by Bevan, Telford was consulted and it is

associated. There were, indeed, very few canal works in England about which Telford was not consulted either in his own right or as engineer to the Exchequer Loan Bill Commissioners.

In Scotland, too, he examined and reported upon the Forth & Clyde Canal, surveyed the Edinburgh & Glasgow Union Canal which was built by Baird, and was engineer of the Glasgow, Paisley & Ardrossan Canal which was never completed but, significantly, turned itself into a railway east of Paisley. Perhaps the most important of Telford's canal works in England in these middle years, and certainly the one in which he was most closely concerned, was that carried out for the Weaver Navigation Trustees from 1810 onwards when he succeeded Johnston as their engineer. He completed the Weston Canal which had been begun by Johnston with the object of by-passing the lower tidal channel of the river and he carried out considerable harbour works at Weston Point where the Navigation enters the Mersey estuary.

None of these activities, however, deserve more than a passing mention here, for it was not until nearly the end of his life that he was called upon to carry out in England canal works upon a scale comparable with those on the Ellesmere Canal which had first won him national renown. For it was in 1824 that the last great period of English canal construction began and even the complacent proprietors of old and prosperous navigations belatedly roused themselves from their long lethargy. There were two closely connected reasons for this canal revival. England's economy had at last recovered from the effect of the Napoleonic War and company promoters found that they could once again readily obtain capital backing for their schemes. This fact alone might not have roused the canal interests to action had not rumours of new railway schemes spread like wildfire through the country and found many willing ears in that new, optimistic and speculative commerical world.

The source of this new interest was the Stockton & Darlington Railway which had been authorised in 1821. Horse tramways had become a commonplace by this time, but in almost every case they were looked upon as feeders of traffic to canals and many of them, like the Ruabon

---

possible that the canal's long summit level over the Naseby Wolds represents a part of an older scheme which Telford projected in 1803. This was for a contour canal with an unbroken level of sixty miles in the same area. Telford reported on the Regent's Canal to the Exchequer Loan Bill Commissioners.

Brook Tramway mentioned earlier, had actually been promoted and built by canal companies. The Stockton & Darlington, however, was something different both in origin and extent. It had no connection with any canal interest either financially or physically, and its total length including branches was to be thirty-six miles. But what really made the canal proprietors sit up and take notice was the appointment of George Stephenson as engineer to the railway. It was as a result of his recommendations that the Stockton & Darlington Company went to Parliament again in 1823 and obtained powers to use 'locomotive engines' and to vary their original line in order to make it suitable for locomotive working. Here was a serious, not to say defiant, threat to the canal monopoly.

The canal proprietors attacked this novel adversary by circulating tracts in which ridicule was blended with Awful Warnings designed to freeze the reader's marrow. A typical example entitled *Observations on Railways with Locomotive high-pressure Steam engines* and dated March 1825, has survived among Telford's papers. There are few objections, real or imaginary, which the anonymous author of this pamphlet has not thought up: the demand for iron would raise the price of coal; the iron rails would 'crackle' in frost and sink in flood; serious delays would result from the need of the iron horse to stop frequently for water which, he insists, 'must be boiling'. Of the rails, he explains that 'the trough will be in the wheel and not (as heretofore) will the wheel run in the trough'. For some mysterious reason he deduces that such rails will present a fearful hazard to all horse traffic at level crossings and he paints a dire picture of the consequences: 'He [the horse] and all confided to his care will crack like lobsters under the resistless engine.' But the worst fate he reserves for those foolhardy enough to travel on the new railway. 'We proceed a few miles from London,' he writes, 'and in crossing the Great North Road at the novel railway speed of twelve miles an hour a sudden explosion unkennels the passengers, parboils the pilot and attendants and scatters the luggage all abroad like that of a vaniquished army. ...'

Though they might pour scorn on the railway threat in public, privately the canal proprietors were seriously alarmed. In June 1824 Caldwell, the Trent & Mersey Canal Committee Chairman, wrote an indignant letter to Telford. 'Is *Inland Navigation* to be ruined by these *Railroad Projectors*?' he asked. 'If so Canal Proprietors had better stop all further improvement and then what will soon be the state of the country?' The proposal to build a railway from Birmingham to Merseyside across part of their territory threw the Ellesmere & Chester Canal Company into a state bordering on panic. 'The rail-roaders are very active in the fields,' wrote the Company's agent, Thomas Stanton, from Ellesmere Canal Office to Telford in December 1825. 'I was a good part of last week in Chester urging Mr. Potts and Mr. Humberstone to active exertions against them.' The Company feared the railway would bring Staffordshire coal into the area to threaten their hitherto secure monopoly of supply from the Ruabon field. As a counter measure, Stanton suggested to Telford that they should immediately begin cutting the long-disputed branch canal to Middlewich without waiting for the approval of the Trent & Mersey or Parliamentary sanction to form the hitherto forbidden junction between the two canals at Middlewich.

In these circumstances it is not surprising that Telford's hope of cutting down his commitments as old age came upon him was unfulfilled. With Jessop and Rennie dead it was to Telford as the undisputed head of his profession that the anxious canal proprietors appealed for aid. What were Telford's own views on railways? This is not an easy question to answer because it is difficult to determine the extent to which his expressions of opinion were biased by his sense of loyalty to the many canal companies he served. His experience dated back to his visit to the Peak Forest Tramway in 1797, while the plate tramways laid down during the construction of the Caledonian Canal almost certainly represented the first extensive use of temporary railways on a civil engineering project. In 1810 he personally surveyed the route for a railway 125 miles long between Glasgow and Berwick. It was to consist of two lines of cast-iron rails on stone block sleepers and apart from two inclined planes the ruling gradient was to be no steeper than 1 in 117 or forty-five feet per mile. In his report on the Glasgow & Berwick, Telford expressed himself in favour of railways where the lie of the land was unfavourable to canal cutting or where water supplies were inadequate.[1]

1. He had earlier expressed the same view in his contribution to Plymley's *Agriculture of Shropshire*.

Telford's next connection with railways came about in 1822 when he was asked by the promoters of the Stratford & Moreton Railway[1] to weigh the merits of two rival proposals. A Mr. Thomas Brown of Halesowen recommended an easily graded line suitable for locomotive traction using 150 lb. malleable iron rails 15 feet long on stone block sleepers at 3-feet spacing. This plan had the energetic support of William James, that enthusiastic but unlucky advocate of locomotive railways, who was Deputy Chairman of the Managing Committee and who saw the Stratford & Moreton as the first link in a trunk route to London. John Greaves, on the other hand, advocated horse haulage with wagons of 2 tons capacity running on cast-iron rails 4 feet long and weighing 36 lb. He planned to use inclined planes with rope haulage, the rest of his line being level. In his report dated November 1822, Telford declared himself in favour of horse haulage and it was in this form that the railway was built by J. U. Rastrick of Stourbridge, Telford acting as consultant.

Because of his association with so many canal projects, railway promoters looked upon Telford as their arch-enemy and did their utmost to discredit his opinions. He found himself in a most invidious position, therefore, when both the Newcastle & Carlisle and the Liverpool & Manchester Railways appealed to the Exchequer Loan Bill Commissioners for financial aid and he, as engineer to the Commissioners, was requested to examine and report upon both projects. In the case of the Newcastle & Carlisle he pronounced in favour of a railway as opposed to a rival scheme for a canal, but he recommended horse haulage and advised against the use of locomotives.

Over the construction of the Liverpool & Manchester Railway there were from the outset such frequent and often bitter clashes of opinion between the engineers and surveyors responsible – William James, the young Rennies, Charles Vignoles, Joseph Locke and the great George Stephenson – that it was with the greatest reluctance that Telford was forced to put his head into such a hornets' nest. But if he hoped to dispose of the matter at a safe distance he was disappointed. The Company failed to supply any of the facts, plans and figures he asked for so that he was compelled to send his assistant, James Mills, to investigate

1. This line ran from the Stratford canal basin at Stratford-on-Avon to Moreton-in-Marsh with a branch to Shipston-on-Stour. The footbridge across the Avon near the Memorial Theatre is a relic of this tramway.

on the spot. From the moment he set foot in Liverpool, the unfortunate Mills found himself ignored by everyone in authority and his only source of information was 'a young man of the name of Gooch,[1] an apprentice of Mr. Stephenson's'. This cavalier treatment of his assistant drove Telford to visit the works himself, and his arrival in Liverpool in January 1829 was an event which even George Stephenson could not ignore. He personally conducted Telford and Mills over the whole line. 'To accomplish this', wrote Telford in his report to the Commissioners, 'upon a line of 30 miles in the present unfinished and complex state of the works was a tedious and laborious task.' It was so indeed for a man of seventy-two in failing health.

Telford discovered that the organisation – or the lack of it – under which the works were being conducted compared very unfavourably with the orderly system of large-scale contracting under the direction of chief and resident engineers which he had been the first to establish. The popular notion that it was the railway builders who first evolved such a system is thus quite wrong. On the state of the works themselves, however, his report was generally favourable and his only serious criticism concerned the fixed haulage engines which were then proposed. He said he was at a loss to know how the railway could be worked effectively by such means. He seems to have gone out of his way to be impartial and to avoid any accusation of anti-railway bias. Yet his report provoked a storm of abuse from the Liverpool & Manchester directors. They published a printed statement in which they described it as: 'an extraordinary document than which one more abounding with inaccuracies and erroneous statements can hardly be conceived'.

This was asking for trouble and the directors soon got it. Telford promptly issued a second report in which he no longer minced his words. He declared that his assistants had been given no help whatever in preparing their surveys and estimates and that costly mistakes in construction must be expected where there existed no proper survey and where the works were being carried on under no proper system. This time he roundly condemned the use of inclined planes and fixed engines, declaring that the Company should never have departed from the Rennies' survey of a level road throughout. The latter, he main-

1. Presumably Tom Gooch as his celebrated brother Daniel was only twelve years old at this time.

tained, would be cheaper both to construct and to operate, using either horses or 'locomotive engines' for haulage over the whole length of the railway.

At this the directors capitulated. As they badly needed money and as Telford held the purse-strings of the Loan Commissioners they must have realised belatedly that they were in no position to ride a high horse. The idea of fixed engines and inclined planes was precipitately abandoned in favour of locomotive haulage and the Company proceeded to set its house in order with such effect that Telford was able to recommend loans totalling £100,000.

Contrary to popular belief, Telford never seems to have questioned the technical practicability of the steam locomotive. He maintained, however, that the advantage of the locomotive over the horse was not great enough to offset either the increased capital cost of rails and bridges capable of carrying engines or the increased cost of track maintenance. In his day this was a perfectly valid argument, but it was shortsighted because it failed to anticipate the very rapid improvement which took place, not only in the locomotive but, more important still in the road it ran upon. Until his death Telford believed that the future of mechanical land transport lay with the steam carriage whose movements were unrestricted and which did not require a costly metal way of its own. He was an enthusiastic advocate of steam road transport and supported a scheme for the introduction of steam carriages on his Holyhead road. It was with this end in view that, notwithstanding his advanced age and ill health, he essayed a journey from London to Birmingham in Sir Charles Dance's steam carriage in October 1831. Boiler trouble at Stony Stratford, fifty-four miles from London, put an end to the attempt, but Telford and the friends who accompanied him, Rickman, Colonel Pasley, McNeill the Holyhead road engineer, Bryan Donkin, Field and Bramah, all seem to have been favourably impressed. But shortly after this the Government empowered the Turnpike Trusts to kill the road steamer by punitive tolls and it was soon submerged by the rising tide of railway development.

Notes written by Telford in 1824 show that he realised sooner than many railway promoters did that on a locomotive-worked railway all the traffic would have to be handled by the Company who owned the line. In his eyes this was another weighty argument against the locomotive and in favour of the steam road carriage. For it was accepted

without question at this time that any form of road, whether it was a
waterway, a turnpike or a railway should be freely open to anyone who
cared to use it on payment of the statutory tolls. The idea that any
company of proprietors should be empowered to build a special road
across England upon which they could claim a traffic monopoly was
an entirely novel, and to many people a repugnant, one. The strength
of feeling on this question may be judged by the fact that canal companies were forbidden by law to trade on their own canals.

Notwithstanding Telford's doubts about the future of locomotive
railways he took pains to keep himself fully informed on the subject.
In January 1824 he dispatched an assistant, H. R. Palmer, to Darlington
with instructions to examine and report to him on the works of the
Stockton & Darlington Railway and then to visit Stephenson's Hetton
Colliery Railway and study the performance of the locomotives already
at work there. Working under Telford's direction, Palmer carried out
a series of comparative experiments on the resistance to traction of the
locomotives and horses working on the Hetton Railway and the boats
on the Ellesmere Canal. This was the prelude to a whole series of
experiments which were extended to the Mersey & Irwell and Grand
Junction Canals and in which Bryan Donkin, Peter Barlow, Bevan and
Chapman collaborated. Their conclusion was that: 'with small
velocities the force of traction on Canals is less than on railways, and
when the velocity is equal to four miles an hour, the forces are equal.
Beyond this velocity the advantage is in favour of the railways.' In the
last two years of his life Telford carried out further experiments in this
field himself at the Adelaide Gallery[1] in London, using a model canal 70
feet long and 4 feet wide and model boats, some of them propelled by
clockwork. One of Palmer's reports on his Ellesmere Canal experiments
throws an interesting light on canal working conditions which were
apparently taken for granted at this time. He found that a 'common
barge' with a load of twenty-one tons had been hauled sixteen miles a
day for six weeks by one boatman and his thirteen-year-old daughter.
Average speed was one mile six furlongs per hour and the estimated
tractive effort fifty pounds.

In 1829, when the Stockton & Darlington had been working four
years, G. H. Buck made a further report to Telford on the railway the
contents of which, if they were so divulged, must have lightened the

1. On the site of the present National Gallery.

hearts of the canal proprietors. Buck concluded that as both loco-motives and horses were being used there could not be much to choose between them. Of the locomotives he writes: 'One [obviously Timothy Hackworth's famous *Royal George*] most approved is a 12 horsepower and weights 12 tons. It moves on 6 wheels and generally conducts 28 waggons containing 74 Tons, moves with a velocity of 5 miles an hour, in some parts 9 or 10. The other engines move on 4 wheels, weigh 8 tons and are each 8 horsepower, drawing 20 waggons with 53 tons at the same velocity. The 12 horsepower engine has taken 36 waggons but this depends upon the wind and the state of the rails. One horse draws 4 waggons, 10 tons 12 cwt, at 3 miles an hour. The horses draw more per horsepower than the engines.' Of the railway Buck has little good to say. The embankments were not built wide enough and were bulging under the weight of the engines. The rails were too weak for their loads and the stone block sleepers too small and of bad material. He condemned the method of fixing the chairs to the blocks by means of spikes driven into wooden plugs set in the stone. He found that either the spikes split the stone when they were driven in or the wooden plugs shrank and worked loose. The whole railway, he concluded, was going rapidly to decay. One suspects that Buck was himself one of the canal party and so looked diligently for the worst; yet it was in truth touch and go for the railway pioneers just at this time and it was probably the performance of the *Royal George*, more than any other single factor, which tipped the scales in their favour.

While experiments were made and rumours and reports flew to and fro, Telford found himself generalissimo of the canal companies' belated campaign to keep the iron enemy out of the Midland shires where they had lorded it so long. Even when they were faced by a common peril the old inter-company feuds were not forgotten so that there was no truly concerted action, but two of the most powerful and wealthy of the old companies, the Trent & Mersey and the Birming-ham, took the field.

In the history of the canals the Trent & Mersey Company occupies a position very similar to that of the London & North Western Com-pany in railway history. In their respective spheres both were entitled to boast that they owned the premier trunk-line and the effect of this distinction upon their behaviour was very much the same. Both com-bined a policy of rigid conservatism with an attitude towards neighbouring companies which was at the best condescending and at the worst bitterly and unscrupulously hostile. Although the canal carried an enormous traffic and the Company was extremely prosperous no substantial improvement had been made to Brindley's 'Grand Trunk' since it had been built. From a traffic point of view by far the worst feature of the canal was the great tunnel under Harecastle Hill on the summit level. Only a single boat's width and nearly 3,000 yards long, it had been a serious bottleneck ever since it was opened and in recent years it had suffered badly from wear and tear and mining subsidence. While one boatman, poised perilously between foul air and foul water, slowly and laboriously 'legged' his way through the darkness of a tunnel no larger than a medieval sewer, his fellows behind quarrelled and fought for the privilege of next turn through while at the other end a queue of boats waited impatiently for the opposing traffic to clear. Harecastle was the subject of incessant complaints from traders, but the Company turned a deaf ear to them until 1820 when they called in John Rennie to inspect the tunnel.

Rennie's report revealed an appalling state of affairs. In some places the tunnel roof was only six feet above water-level and in others it was so narrow that the brick lining, only nine inches thick at best, had been worn to half its thickness by the constant rubbing of the boats. Where the tunnel walls were wet the mortar had become as soft as clay so that bricks could be pulled out by hand. In the branch tunnels which Brindley had driven to communicate directly with the workings of the Goldenhill Colliery, conditions were even worse. The arches at their junctions with the main tunnel had been 'torn to pieces' by the action of boats turning in or out of them. No attempt was made to keep their channels clear of mud and rubbish from the coal workings and this was carried along them into the main canal. Without giving any notice to the Company, the miners were in the habit of breaking into these branch tunnels wherever it happened to be most convenient for them to load their coal into boats. 'Such places', concluded Rennie, 'can scarcely be called safe.' There was only one solution, said Rennie: the tunnel must be closed to all traffic for at least twelve months so that it could be enlarged and completely repaired. He put forward three proposals for passing the traffic while this work was done:

(1) To carry a temporary line of railway over Harecastle Hill as had

been done by the Grand Junction Company while their Blisworth tunnel was building.

(2) To build a by-pass canal through the Bath Pool valley which would involve a deep cutting and a chain of sixteen locks.

(3) To drive a second parallel tunnel.

Of these three choices Rennie recommended the tunnel and estimated the cost at £58,000. But nothing was done until after Rennie's death when, in January 1822, Telford was invited to assist.

Telford's visit to Harecastle was delayed by one of his usual protracted tours of inspection, but on the 4th of March he finally arrived at The Red Bull at Lawton near the north end of the tunnel where he was met by two representatives of the Company, Heath and Johnson, and a young man named James Potter who had just been appointed resident engineer at Harecastle. The Company were by this time not only concerned about the tunnel but about the water supply to their summit level and Telford's visit had a fourfold object: to survey the existing tunnel; to determine the site of a new tunnel; to examine the river Dane feeder and to survey the site of a proposed new reservoir in a valley near Knipersley a few miles to the east. He went through the tunnel in a boat with Potter. Perhaps some first-aid measures had been taken since Rennie's visit for he pronounced its condition to be 'tolerably good' and saw 'no symptoms of immediate danger or failure'. He recommended that a new tunnel should be driven at a distance of twenty-five yards from the old one and he estimated the cost at £60,000. He also recommended that James Potter should be appointed engineer in charge of the works at a salary of £500 a year as although he was 'a very young man' he had formed the highest opinion of his ability.

'The canal revenue being about £140,000 a year, all these improvements can be very well afforded,' wrote Telford subsequently in a letter to Count von Platen, yet still the Trent & Mersey Company hesitated to make a move and for two more years nothing was done. Ultimately, however, the defeatist attitude to the railway threat which Chairman Caldwell expressed to Telford did not prevail and in the spring of 1824 tenders for the new tunnel were invited. It was the prelude to one of the most remarkable tunnelling feats in engineering history.

The successful tender was submitted by the firm of Pritchard &

Hoof of Kings Norton and Daniel Pritchard, in his letter to Telford covering his contract estimates, dispels the notion that large-scale civil engineering contracting was a product of the railway building age. For Pritchard states that his firm had been responsible for driving the following canal tunnels and he quotes the cost per yard in labour in each case: Bosworth and Crick tunnels on the Grand Union Canal; Maida Hill and Islington on the Regent's Canal; Kendal tunnel on the Lancaster Canal, and finally the great Thames & Medway tunnel near Rochester which subsequently became a railway tunnel. Of Harecastle Pritchard wrote: 'The Rock I find to be extremely hard, some of it in my opinion is much harder than ever any tunnel has been driven in before excepting the one that is executed by the side of it.' Coming from such an experienced man, this remark goes a long way towards explaining why Brindley's navigators toiled for eleven years to drive the old tunnel.

Pritchard's estimate reads as follows: 'For excavating the ground for the Tunnel, putting in necessary timbers, winding the spoil to the surface, loading into skips and lowering the bricks, mortar and other materials, bricking the tunnel with such strength of work as the Resident Engineer may think proper, setting and taking down the centres, drawing the timbers and pumping the water, per yard £14.' This was for labour only. In addition he was to: 'find all materials, Tools, Gins, ropes, skips, iron railways for the underground work. The Company to find bricks, lime, sand, coals, Timber, centres and laggings and to convey them to the Pit Bank and likewise to keep the Tunnel clear of water.'

Having hesitated for so long the Company now seem to have decided that the tunnel must be built with the utmost speed almost regardless of cost. To this end no less than fifteen shafts were sunk down to tunnel level from the top of Harecastle Hill, ten of which were equipped with horse gins. In addition sixteen cross headings were driven to communicate with the old tunnel. This multiplicity of shafts and headings far exceeded previous practice in tunnel boring and increased the cost, but the work was thereby greatly accelerated because it could proceed simultaneously at so many different points. Boulton & Watt beam pumping engines were installed at each end of the tunnel to keep the workings clear of water, the larger at the northern end being referred to as the Nelson engine. Another smaller steam engine which had originally been used as a pump at Lawton Locks was moved to Long-port at the south end of the tunnel where it was adapted to drive a clay mill for the brickyards which were opened there to supply the the needs of the contractors. Special lime for mortar was shipped from Barrow-in-Furness and Potter bought in Liverpool a special mortar mill which had been used for dock construction. Railways were laid from the brickyards to the tunnel and from the north end cutting over the hill via an inclined plane. The latter raised spoil, stores and also coal for the Nelson engine. The presence of the old tunnel undoubtedly speeded up operations for at night when the traffic ceased boats could be run to the cross headings where they loaded spoil or discharged bricks, mortar and centering timbers.

Harecastle Hill must have presented a dramatic spectacle during this brief period of hectic activity when the canals were making their last fruitless bid to retain their supremacy against the threat of steam power and the iron rail. Barrack buildings and stables had been built on the hill so that fresh shifts of men and horses could enable work to go on continuously. Night and day the pumping engines, belching smoke and steam, maintained their ponderous but unfaltering rhythm, a hoarse sigh at each slow stroke punctuated by the syncopated clatter of the valve gear trip levers. Night and day horse gins creaked, trucks rumbled along their iron ways and the furnaces of the brick kilns glowed red.

The northern end of the tunnel gave Pritchard & Hoof the most trouble. First their miners encountered a water-bearing quicksand which taxed the power of the Nelson engine to the utmost and then, having mastered this, they had to hack and blast their way through the band of pitilessly hard Millstone grit and Rowley Rag which had caused Brindley's navigators so much trouble. Yet the rate of progress was fantastic. The work of sinking the fifteen shafts had begun in the summer of 1824 and just a year later Potter was able to report to Telford that all had been sunk to depth and that excavation was going on in nineteen different places. By October of 1825 a heading seven feet in diameter had been opened throughout and a considerable portion had been completed and lined. On 12 June 1826 Potter forecast completion on the following 1st of June. This must have seemed mere youthful optimism on his part, yet in fact the contractors beat his estimated schedule by a comfortable margin, for on 1 March 1827

Potter was able to write to Telford as follows: 'The general stoppage for repairing the canal will take place on the 8th of April during which we shall take out the Dams at each end of the Tunnel and open it to the Trade on the 16th.'

So, in less than three years from the first stroke of a spade on Harecastle Hill, there was completed a tunnel 3,000 yards long, and fourteen feet in diameter, lined with seven million bricks and including a towing path on masonry arches. It could be said that the tunnel itself was built in under two years, for the first brick was laid on 21 February 1825 and the last on 25 November 1826, the remainder of the time being spent in building the towpath. 'It is', wrote Telford, 'the most perfect work of its kind yet executed.' He had good reason to be proud, for the whole operation had been carried out in exact accordance with the plans set out in the memorandum which he had dated from Longport on 7 July 1824. Moreover, notwithstanding the speed of the work, true to the Telford tradition, not a single life was lost. This, added Telford, was 'unexampled in such an undertaking'. It was indeed, and long would it remain so. The driving of the Harecastle New Tunnel was a feat which the rising generation of railway engineers, for all their brilliance, were never able to surpass. It was a great achievement, too, on the part of the contractors, Pritchard & Hoof, an achievement which ranks equal to that of their forerunners who had driven the first major tunnel in the world under the same hill. With the exception of the southern approach cutting for which Messrs. Dutton & Buckley, the contractors for Knipersley Reservoir, were responsible, the whole of the work had been in their hands.

Like Telford's earlier tunnels at Chirk and Whitehouses, upon which it was modelled, the new Harecastle was a one-way tunnel. This would have been no disadvantage if Telford had been able to complete his plans for, as he told von Platen, he proposed to enlarge the old tunnel and build a towpath through it also. But the Trent & Mersey Company would not agree to the additional expenditure with the consequence that southbound traffic still had to be legged laboriously through the old tunnel. Both tunnels continued to be used in this way for many years until the old one became unsafe and all boats were worked by tug through the new. On Telford's recommendation the use of the old branch tunnels to the coal workings was discontinued when his new tunnel was built.

An ironical footnote to this success story is that the building of the Knipersley Reservoir which had seemed by comparison with the tunnel a simple and straightforward task caused Telford a great deal of difficulty and anxiety. There was continual trouble with leakage through the discharge cock or through the masonry of the well in which it was situated, and through the outlet culvert or the dam itself. At times it was serious enough to alarm the owners of land in the valley below the dam, and the reservoir became the subject of much bitter dissension at the meetings of the Trent & Mersey Special Committee at Stone. In November 1828 the Committee consulted two local men about the reservoir, much to the annoyance of Telford and Potter. One was George Hamilton, the church architect of Stone and the other the more noted architect and civil engineer, James Trubshaw. 'This is, of course, not only an insult to me but also to you,' wrote Potter to Telford, and of Hamilton he added 'like the old man in the story he tried to please everyone and succeeded in pleasing no one.' The employment of Trubshaw must have been particularly galling to Telford for the former was at that time building, and later completed successfully, the Grosvenor Bridge over the Dee at Chester to a design by Thomas Harrison which Telford had pronounced impracticable. In December Potter announced his intention of leaving the Company. He had received no pay, he said, since the completion of the tunnel and with the exception of Josiah Wedgwood, whom he described as 'the most intelligent and gentlemanly of the whole set', his opinion of the Trent & Mersey Committee appears to have been very low indeed. In May 1829, Trubshaw announced that the reservoir, which had been emptied as a precautionary measure in the previous July, was ready for filling. It must have been with a certain malicious satisfaction that Telford read a letter from William Vaughan, the Trent & Mersey secretary, written in the following September, complaining that Knipersley Reservoir was leaking as badly as ever and appealing for his help. If Telford had any further correspondence on the subject it does not appear to have survived, so that in the absence of the Trent & Mersey minute books it is impossible to say when the trouble was finally mastered.

Closely connected with these Trent & Mersey improvements and another move in the campaign against the railway threat was the promotion of the new Macclesfield Canal on a line surveyed by Telford early in 1825. He made his final report on the 1st of December in that year and the Company's Act received the Royal Assent in April 1827. Turning northwards from the Trent & Mersey at Hardings Wood near the north end of Harecastle Tunnel and running for $27\frac{1}{2}$ miles to a junction with the Peak Forest Canal at Marple, the new waterway not only served the growing towns of Congleton and Macclesfield which had hitherto been isolated from the canal system but it also opened up a much more direct water route between the Potteries and Manchester. Telford's survey gave the canal two long levels, the only locks being the flight of twelve at Bosley, situated between Congleton and Macclesfield. Characteristically, the Trent & Mersey Company only grudgingly conceded this newcomer's right to exist after the most stringent water restrictions had been imposed upon it. In no circumstances might a drop of the sacred waters of the Trent & Mersey be permitted to escape into the new waterway. The Macclesfield Company were not even permitted to make the junction between the two canals. Instead, Pritchard & Hoof cut for the Trent & Mersey a branch canal one and half miles long to connect with the Macclesfield through a tedious arrangement of two stop locks, one built by each company. Conversely the Trent & Mersey secured the right to receive all the water flowing down the Macclesfield from that Company's five storage reservoirs at Pot Shrigley, Sutton and Bosley.

Thomas Brown of Manchester was the engineer of the Macclesfield Canal and Telford appears to have played little or no part in the actual construction even in a consulting capacity, although both he and William Provis took up shares in the Company. As we shall see presently, both men were fully occupied elsewhere at this time. Telford's line ran high above the great plain of Cheshire as it followed the western slopes of 'the backbone of England'. But although it clung to the contours around the 400-feet level the works were heavy, for numerous embankments and aqueducts were required to carry the canal over the valleys of streams falling into the plain.

While the Trent & Mersey Company were hovering on the brink of launching their new tunnel scheme Telford was inspecting the Birmingham Canal with a view to making improvements of a similar kind. The Company of Proprietors of the Birmingham Canal Navigations wielded a power in the canal world equal to that of the Trent & Mersey and the Bridgewater Companies. They controlled a network of waterways in the Black Country which formed the hub of the

Midlands canal system. This had grown up around Brindley's original canal from Birmingham to his Staffordshire & Worcestershire Canal at Aldersley Junction which had received its Act in 1768. It was still the main artery of the system and the traffic upon it was tremendous. In a letter to Count von Platen written soon after his first inspection, Telford said it was no uncommon thing for two hundred boats to pass through a lock within twenty-four hours. In the same letter Telford estimated the Company's income at between £80,000 and £90,000 a year. Yet for nearly forty years nothing had been done to improve Brindley's canal and its condition is best described in Telford's own words in his autobiography.

'Upon inspection,' he writes, 'I found adjacent to this great and flourishing town [Birmingham] a canal little better than a crooked ditch with scarcely the appearance of a haling-path, the horses frequently sliding and staggering in the water, the haling-lines sweeping the gravel into the canal and the entanglement at the meeting of the boats incessant; while at the locks at each end of the short summit crowds of boatmen were always quarrelling, or offering premiums for a preference of passage, and the mine owners, injured by the delay, were loud in their just complaints.'

The short summit level to which Telford refers was at Smethwick. As built by Brindley it was only one mile in length, but by 1787 it had been lowered eighteen feet. This reduced the number of locks, but the short summit still remained and was dependent for its water supply on the continuous efforts of pumping engines which lifted water out of mine workings. A more inconvenient, expensive and precarious method of operating a busy canal could not be conceived, and now at last, faced by the threat of railways, the Birmingham Company resolved to do something about it. In his preliminary report to John Freeth, the Company's secretary, Telford advised them to start work on the section between Birmingham and Wednesbury as this carried the most traffic and because it was 'most complicated and imperfect as regards the working of locks, the supply of water and the numerous turnings in the canal itself'. 'When it is considered', Telford went on, 'that at Smethwick there are six locks and two steam engines, at Spon Lane three locks, and that from the Wednesbury level the water has to fall down uselessly to the lowest level and be from thence pumped up to the summit by the Ockerhill Engines, I shall be fully justified in recommending the committee to excavate a new canal along the southern side of the present one between Smethwick and Spon Lane so as to bring the Birmingham and Wednesbury levels upon an equality.'

This is precisely what was done. Under Telford's direction a mighty cutting seventy feet deep was driven through the old Smethwick summit and within it ran a new canal forty feet wide with a broad towpath on each side. Several roads were carried over the canal at high level, Galton Bridge with its single lofty cast-iron span being the most notable. Everywhere the bridges were built to span the new canal at full width in order that canal traffic should not be hindered by narrow 'bridge holes'. This was the work which the younger Brunel described as 'prodigious' when he visited Birmingham in 1830. Straight and wide, Telford's canal cut its way through a country torn asunder by coal and iron workings which in his day flared and fumed with furnaces and ever-burning gob fires but which has now become a cold desert of ash and cinder and of desolate pools like the drowned craters of dead volcanoes. It reduced the canal distance between Birmingham and Wolverhampton by no less than eight miles and Brindley's old tortuous navigation survived only as a series of loops and branches winding away to right and left of the new straight cut. At Rotton Park a great reservoir was built to feed the new canal, while further west, between Deepfield and Bloomfield, there was more deep cutting and the new Coseley Tunnel which was built wide enough for two-way traffic and two towing paths. The only part of Telford's plan which was not carried out was the widening of the three locks at Tipton. Like the old tunnel at Harecastle they survived as a monument to the ill-judged parsimony of the canal companies even in their hour of peril.

'It was commonly said in Birmingham', wrote John Rickman after his friend's death, 'that Mr. Telford ought to have had a public reward for introducing good manners among the boatmen, who seldom passed each other without quarrels and imprecations arising from the difficulty and delay of passing the towing line under the inner boat whereas they now meet and pass in good humour and with mutual salutations.' Telford was indeed a friend to the canal boatman. When he returned to inspect the new tunnel at Harecastle in 1829 he asked a passing boatman how he liked it and took great delight in the man's reply that he only wished it extended all the way to Manchester.

South-east of Birmingham, Telford planned another canal improvement upon an even grander scale which, had it come to fruition would have placed the canals in a far stronger position to meet the competition of the London & Birmingham Railway. This was the London & Birmingham Junction canal scheme, a new waterway which was designed to run from the Oxford Canal to the Stratford Canal, crossing what is now the main line of Grand Union Canal near Solihull and so avoiding the tedious lockage into and out of the Avon valley at Warwick which mars the London to Birmingham canal route to this day. For this new link Telford planned a magnificently wide waterway with two towing paths and paired wide locks. Had it been built, the north end of the Oxford Canal between Braunston and Brinklow would have become part of the London to Birmingham route and it was in anticipation of this that the Oxford Canal Company built a costly series of new cut-offs which shortened Brindley's old canal between these two places by no less than thirteen miles.[1] But these hopes were never fulfilled. The promotion appears to have received no genuine backing but to have been a highly speculative political move which was decisively rejected in 1830. Though subsequently revived in varying forms it never materialised.

North of the Black Country, Telford's plans met with more success. Here he was commissioned by the Birmingham Canal Company to survey a line for a new direct canal route from Wolverhampton to the Mersey in opposition to the Birmingham & Liverpool Railway scheme which was causing the canal proprietors so much anxiety. Telford went quickly to work and Thomas Eyre Lee, W. R. Palmer and James Mills were all employed by him on surveys and calculations which resulted in an estimate of £388,000 for building a new canal which would link the Staffordshire & Worcestershire at Autherley, near Wolverhampton, with the Ellesmere & Chester Canal at Nantwich by the most direct route. By this means the railway scheme was repulsed

1. The London & Birmingham Junction was at first projected to run to Braunston, a plan which would have excluded the Oxford Canal altogether. It was probably this threat which forced the latter to straighten their line and thus to induce the London & Birmingham to alter and shorten their projected canal so as to bring the Oxford a share of the traffic.

and the Birmingham & Liverpool Junction Canal Company was incorporated in May 1826. It was destined to be the last victory of the canals over the all-conquering railway, the last trunk canal route to be built in England and Telford's last great engineering work. He forecast completion in four years, but he was to die before the canal was finished.

II

THE LAST CANAL

IF a man were to set out to walk from Wolverhampton to Merseyside he could not do better than follow the towpath of Telford's Birmingham & Liverpool Junction Canal. This fact emphasises the contrast between this, the latest of the English canals, and those which preceded it. Although towpath walking is a delightful pastime it is seldom or never the most direct way of getting from one place to another. In this case, however, it is quite otherwise. A glance at a map shows that the directness of Telford's canal is unmatched by either road or rail. It was not that the lie of this rolling country on the Shropshire–Staffordshire borders favoured such a course; it was that the engineer abandoned preconceived notions of canal cutting.

The difference between Brindley's winding contour canals and the long straight levels of the Birmingham & Liverpool Junction with their great cuttings and embankments is evidence of more than a rapid development of civil engineering technique. It is a manifestation of the profound change which had taken place in the social and economic structure of England within the compass of Telford's long lifespan. He lived through the great agricultural enclosure movement which transformed the texture of the English rural landscape and he saw groups of villages grow as dispossessed countrymen turned to domestic industry until new factories sprang up to absorb them and the villages coalesced into sprawling, smoke-laden industrial towns. Although he could not fully appreciate the significance of these changes, he was witnessing the death of the old rural society and the birth pangs of a new industrial and commercial economy the like of which the world had never seen before. Although Brindley's canals were one of the chief factors in bringing the new order to birth, they had been conceived in terms of the old. Their devious windings were not enforced simply by limited technical resources: Brindley argued that if more villages could be brought within easy reach of his canals by such means the better it would be for them and for the canal companies. In other words he designed his canals to serve purely local ends by bringing the products of industry and the town to the countryman and receiving his produce in exchange. He could not conceive how rapidly the towns would grow once he had given them assured life-lines nor how quickly they would draw the countryman within their orbit. In fact the embryo towns of Brindley's day grew so fast that before the end of Telford's life the navel strings of the old canals could no longer nourish them and so effectually prevented further growth. The new centralised urban economy clamoured for longer and more efficient communications to enable it to expand. The new railway schemes which threatened the handsome dividends of the canal proprietors were not, as so many of them supposed, merely the speculative bubbles blown by a buoyant economy; they were an attempt to find an answer to this insatiable demand. Another decade would prove how successful an answer it was, but meanwhile Telford's Birmingham & Liverpool Junction Canal represented the canal's last belated effort to meet that demand. Had Telford realised it, he was already fighting in a lost cause but this does not alter the fact that his canal was a magnificent feat of civil engineering comparable in magnitude with the works of Stephenson, Brunel and Locke on the first main railway lines.

Typical of the inability of the canal companies to forget their petty squabbles and combine in face of a common enemy was the reaction of the Trent & Mersey to the new Birmingham & Liverpool Junction Company. From Great Haywood Junction westwards the Trent & Mersey formed part of the old roundabout and heavily locked canal route from the Black Country to Merseyside and Manchester. Nevertheless, one might have supposed that the Trent & Mersey Company would have regarded the new canal as the lesser of two evils. For if it and the long-disputed branch to Middlewich were built it would mean that the Trent & Mersey would at least handle all the new route's Manchester traffic between Middlewich and Preston Brook, whereas if the threatened railway had won the day it would in all probability have lost the traffic altogether. But no, the Trent & Mersey reacted in the traditional manner; the new canal threatened its time-honoured trade and must therefore be opposed. Telford's appointment as engineer to the Birmingham & Liverpool was quickly followed by a stilted and pompous letter from Caldwell of the Trent & Mersey

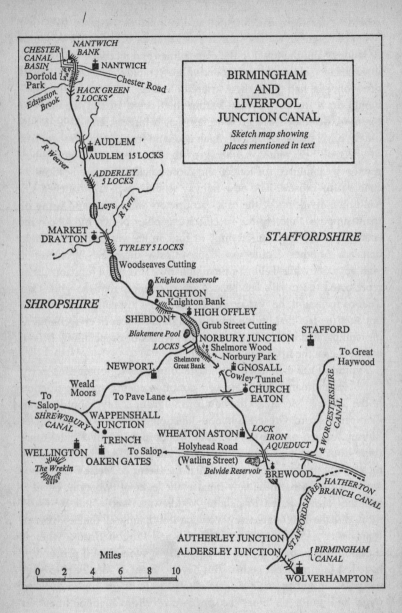

BIRMINGHAM
AND
LIVERPOOL
JUNCTION CANAL

*Sketch map showing
places mentioned in text*

CHESTER
CANAL
BASIN
NANTWICH
BANK
NANTWICH
Dorfold
Park
Chester Road
Edstaston Brook
HACK GREEN
2 LOCKS

R. Weaver
AUDLEM
AUDLEM 15 LOCKS
ADDERLEY
5 LOCKS
Leys
R. Tern
MARKET
DRAYTON
STAFFORDSHIRE
TYRLEY 5 LOCKS
Woodseaves Cutting
Knighton Reservoir
KNIGHTON
Knighton Bank
SHROPSHIRE
SHEBDON
HIGH OFFLEY
Grub Street Cutting
STAFFORD
Blakemere Pool
NORBURY JUNCTION
LOCKS
Shelmore Wood
To Great
Haywood
Norbury Park
NEWPORT
Shelmore
Great Bank
GNOSALL
Cowley Tunnel
Weald
Moors
To Pave Lane
CHURCH
EATON
To
Salop
LOCK
SHREWSBURY
CANAL
WAPPENSHALL
JUNCTION
WHEATON ASTON
IRON
AQUEDUCT
TRENCH
To Salop
Holyhead Road
WELLINGTON
OAKEN GATES
(Watling Street)
STAFFORDSHIRE & WORCESTERSHIRE CANAL
The Wrekin
Belvide Reservoir
BREWOOD
HATHERTON
BRANCH CANAL

AUTHERLEY JUNCTION
ALDERSLEY JUNCTION
BIRMINGHAM
CANAL
Miles
0 2 4 6 8 10
WOLVERHAMPTON

expressing pained surprise at the fact that the engineer of their new tunnel should accept office in a rival concern. But Telford was too old and too eminent a man to be browbeaten in this fashion. He had recently refused an invitation to act as engineer to the new Liverpool & Manchester Railway project partly on the score of his advancing years, but partly also, as he subsequently admitted, out of loyalty to the canal interests he had served so long. But this was carrying things too far and he sent a reply to Caldwell couched in the same vein of stilted dignity. He had already heard, he said, that his work for the Birmingham & Liverpool Junction Canal project had not met with the Company's approbation, and he went on: 'Altho' I considered myself honoured by being selected by a great Company to conduct a difficult work, I never understood that it could thereby be inferred that I was to be so exclusively attached to this service as to interfere with previous engagements or be in any measure precluded from attending to and promoting other improvements. . . .

'Under these circumstances, if the Company should continue to disapprove of the Project mentioned in the beginning of this letter, I am aware of the impropriety of my being in their confidence in regard to any measures they may be disposed to adopt and therefore, altho' I have no wish to decline the management of the Tunnel Works, I shall await the decision of the Select Committee previous to taking any further active steps respecting them.'

This was tantamount to saying that the Trent & Mersey could like it or lump it so far as he was concerned, and the fact that one of the most powerful Companies in the country was thereby compelled to swallow its pride and accept him on his own terms gives us the measure of Telford's stature in the engineering world. In this particular instance Telford's strength lay in the fact that Caldwell's reaction to his letter was a matter of small concern to him. Forced to choose between the two undertakings there can be little doubt where that choice would have lain. Although the new tunnel at Harecastle was engineered in the grand manner, Telford was, after all, only duplicating the work of his great predecessor whereas the Birmingham & Liverpool Junction Canal undertaking possessed the magnitude and the novelty which, as he had told von Platen, he was unable to resist. Perhaps an element of sentiment entered into his calculations also, for this, his last great work, would be closely linked with his first. Indeed, after his death the two canal companies would amalgamate as the Shropshire Union Railways & Canal Company and thus incorporate in one system the alpha and omega of Telford's engineering career.

For the building of the Birmingham & Liverpool Telford called about him for the last time that company of engineers whom he had trained to his methods and who had served him so long. Their presence alone is enough to indicate the importance which he attached to the venture. Their ranks were thinning now. The years had taken David-son, the friend of his youth and that 'treasure of talents' John Simpson. But there was still John Wilson, who remembered the great days of Pont Cysyllte, and his two stalwart sons; still Alexander Easton of the Caledonian and William Provis of the Holyhead road. Easton was appointed resident engineer with headquarters at Market Drayton and with his younger brother George as his assistant and clerk, while Wilson and Provis shared the contract lots between them.

The new canal encountered stiff opposition from landowners along the route. They did not prevail although, in addition to legal and parliamentary expenses, the Company expended no less than £96,119 in compensation to them. The reason why the late canals, like the railways which followed them, ran into this difficulty whereas Brind-ley's canals met with far less opposition, is to be found in the effects of the enclosure movement. The agricultural enclosures ushered in the era of 'high farming' and enormously enhanced the wealth and power of the landed gentry. Where Brindley's surveyors had encountered only common fields and village 'wastes' or at the worst properties of small extent which could, if necessary, be easily avoided, their nine-teenth-century successors found the farmlands, the strictly preserved game coverts and the great landscaped parks of the new squirearchy straddling their lines of route. Such was the opposition which not only succeeded in exacting from the Birmingham & Liverpool such a heavy toll but which forced Telford, with great reluctance, to make two important deviations from his surveyed route in the vicinity of Nantwich and Norbury for which the Company was destined to pay very dearly.

At a meeting which Telford attended at the Lion Hotel at Newport in July 1826, it was agreed to divide the canal for contract purposes into three sections. First, from Nantwich Basin, where the old Chester Canal terminated, to the top of the five locks which were to be built at Tyrley, near Market Drayton; second, from Tyrley through Norbury to the point where the road from Church Eaton to Pave Lane crossed the line of route; third, from this point to the junction with the Staffordshire & Worcestershire Canal at Autherley. The second of these divisions included a branch canal to Newport, falling away by a flight of locks from its junction with the main line at Norbury and extending beyond Newport to meet the old Shrewsbury Canal at Wappenshall. Thus the canal on which Telford had built his first iron aqueduct would at last be linked to the rest of the canal system and the Shropshire coal and iron district likewise. The central division of the main line would have to act as a summit level for this branch and in order to keep it supplied the works included a catchment reservoir in a valley between the villages of Knighton and Great Soudley. A single lock at Wheaton Aston on the third division would raise the canal to the level of the Staffordshire & Worcestershire nearly eight miles away and to keep this top level supplied there was to be a second reservoir at Belvide where the Holyhead road crossed the route. Apart from the solitary exception at Wheaton Aston, Telford had concentrated all the locks into the first and most northerly division; first, two locks at Hack Green near Nantwich, then a flight of fifteen at Audlem which lifted the canal out of the Cheshire plain and finally two flights of five each at Adderley and Tyrley to the north and south of Market Drayton. These three divisions split the thirty-nine miles of the main canal into approximately equal parts and it was decided that work should commence at the northern end.

At a meeting in London on 2 December 1826 the canal committee decided to accept John Wilson's tender of £198,100 for the first division and in January 1827 he began work at Nantwich. It was here that Telford had been forced to make the first of the two important deviations which have been mentioned. The canal had originally been planned to make an end-on junction with the terminal basin of the Chester Canal, but this line would have cut across the corner of Dorfold Park and Mr. Tomkinson of Dorfold Hall raised such strenuous objections that the Company was forced to agree to carry the canal further to the east. The lie of the land was such that this deviation involved nearly half a mile of high, curving embankment, an iron aqueduct over the road from Nantwich to Chester and the diversion of part of a public road known as Henhull Lane. It was closely followed by another high bank over the valley of the Edleston Brook. Other major works in the division were the embankments and aqueducts which carried the canal over the river Weaver at Audlem and the Tern at Market Drayton and the deep cutting at Adderley Leys.

Yet neither these works nor the many locks which had to be built were comparable in magnitude with the task which confronted William Provis when he contracted for the second division in May 1829. Starting from the Tyrley end there was an immense cutting to be driven at Woodseaves, a mile in length and of a maximum depth of ninety feet. Between the villages of Knighton and Shebdon there was to be an embankment a mile long and fifty feet high and near High Offley another deep cutting known as Grub Street nearly two miles in length. This brought the canal to Norbury where the Newport Branch would commence. Just to the south of the site of this junction the second of the deviations began. This was made on the insistence of Lord Anson of Norbury Park who refused to allow the canal to cut through his game preserves in Shelmore Wood. Through this wood the canal could have been driven on a level, but so that his lordship's pheasants might not be disturbed Telford was forced to swing his line westwards in a great arc round the low-lying perimeter of the covert where it had to be carried on an embankment a mile long and sixty feet high. This was the Shelmore Great Bank, as it was called, a name which was to become painfully familiar to the unhappy canal proprietors in the years to come. Two miles south of Shelmore was the last major work on the second division – the 690-yard tunnel at Cowley near Gnosall.

By July 1827 John Wilson had 1,600 men at work on the northern division and the work was proceeding smoothly and rapidly. During this first summer Telford, accompanied by Alexander Easton, carried out two minute inspections of the whole line and expressed himself highly satisfied. In January 1828 he reported that a third of the work had been completed and in the following July he announced that progress had been so rapid that he believed the division should be completed by Michaelmas 1829 instead of April 1830 as had been first forecast. To speed the work, Wilson not only made use of railways but also made a practice of cutting and flooding some of the easier lengths of canal so that boats could be used. For example, by the winter of 1829 seven boats were reported in use carrying spoil from the cutting at

Adderley Leys southwards to the Tern embankment. Yet despite all his efforts Michaelmas passed and the division was not finished, although by the year's end all the masonry work had been completed with the exception of the bottom lock at Audlem. But during that wet winter when the cold rains sluiced down upon the great raw mounds of marl, Telford's troubles began. In January he reported serious slips on the Nantwich embankment. At its base the treacherous marl bulged under the superincumbent weight and spread outwards slowly but with the irresistible pressure of a lava flow to engulf hedges and block Henhull Lane. It would be necessary, wrote Telford, to broaden the base of the embankment, but he advised the Company not to force the work but to allow Wilson to make the bank good in the dry summer season. But the winter was followed by a sodden spring and Nantwich Bank was still on the move. Telford reported in July that Henhull Lane was again blocked and would have to be diverted and that south of the Chester road aqueduct 300 yards of the bank had sunk four feet below canal level. In January 1831 Nantwich Bank was still settling and spreading and fresh spoil was being brought to it by boat. Moreover the next bank at Edleston had also begun to move and 7,000 cubic yards of earth had been dumped at its base in the attempt to check it. By midsummer of this year Telford reported that he had been able to pass through the canal as far south as Norbury but that 300 yards of treacherous embankment at the Nantwich end still tantalisingly isolated it from the waters of the Chester Canal. It was not until a year later that he was able to say that Nantwich Bank was at last complete and secure.

Meanwhile Provis, who had started work at the Tyrley end of the second division in June 1829, was very soon in trouble. In the great cutting at Woodseaves his men encountered strata of treacherous, friable rock alternating with bands of clay which crumbled away when dry and acted as a lubricant when wet. In some places attempts were made to shore up the menacing rock shelves by building dry walls beneath them; in others the dangerous portions were cut back. But notwithstanding these efforts the effect of winter rains and frosts was to dislodge fresh masses of earth and rock and send them thundering down into the bottom of the deep defile. At no small peril to themselves gangs of men had to be regularly employed to clear these slips, not only at Woodseaves but in the long Grub Street cutting where the ground was just as unstable.

In the summer of 1830 Provis began to drive the tunnel at Cowley, beginning at the northern end, but his miners had only advanced ninety yards when they broke into a bad fault. This dangerous place was shored up with heavy timbering and the miners then advanced cautiously for another 150 yards until the rock became so treacherous and rotten that to proceed any further would have been to court disaster. After an urgent conference, Telford, Easton and Provis agreed that the only thing to be done was to open out the remaining 450 yards of tunnel. A year later it was decided to open out a further seventy yards because the rock was disintegrating and this process went on until of the 690-yard bore which had been planned a mere eighty-one yards of canal at the north end remained in tunnel.

In the summer of 1830 Telford announced that the third division of the canal from the Church Eaton bridge to Autherley had been pegged out and the contract drawings and specifications made. John Wilson took up this contract and had a force of 600 men and thirty-five horses at work on it before July was out. Provis contracted for the Newport branch at the same time and for the Knighton reservoir at the end of the year. On 9 January 1831, John Wilson died very suddenly. Telford must have felt this loss very keenly. Fortune which had favoured him so long seemed now, at the last, to have turned against him and when the news was brought to the old engineer his thoughts must have echoed the words of Macbeth: 'He should have died hereafter . . .' True, Wilson's work was to be carried on by his sons but, failing in health since his illness at Cambridge and beset by difficulties such as he had never encountered before in all his long experience, Telford could ill afford to lose the companion of over thirty years who, since the deaths of Davidson and Simpson had become his right-hand man. Notwithstanding the loss of their father, the young Wilsons made excellent progress in the third division, while to the north victory was at last in sight in the long battle with the Nantwich embankment. It must have seemed for a brief time as though the fates had relented and that the troubles had been mastered. But so far as William Provis on the second division was concerned, this was only a brief respite. There were more slips in Woodseaves cutting and on the Knighton embankment, while on the Newport branch he found his new canal sinking into bog where it crossed the Weald Moors on its western extension to Wappenshall. But the worst trouble of all was on Shelmore Great Bank.

Provis had begun work at both ends of the Shelmore embankment in the late summer of 1829. A force of 400 men and seventy horses was concentrated on it. Spoil was brought by wagon to both ends of the bank from the south end of the Grub Street cutting and from the cutting at Gnosall. More was obtained by side cutting, that is to say by lifting the ground on each side of the bank. The bad weather of that winter and the following spring which had started the slips at Nantwich delayed the work at Shelmore, but by the following July 490,000 cubic yards of spoil had been tipped. Next winter the bank had been raised to within a quarter of its full height, but when Telford inspected it he found a state of affairs with which he was by now all too familiar. The bank was slowly subsiding throughout its entire length; already it was 200 feet wide at the base and much of the soil which had been won by side-cutting had returned to the place from which it had been drawn. He decided that it was useless to proceed with the marl got by side-cutting and he ordered this work to cease forthwith. He ruled that in future only the lighter, sandy soils found near each end of the bank should be used and railways were laid to convey this material to the tips. Telford believed that this would bind the treacherous marl and so consolidate the bank and for a time his hopes seemed to be fulfilled. Although settlement still went on and progress was painfully slow, a year later the bank had been raised to its full height for half its length while the remainder was nine feet below. Telford forecast completion by Michaelmas 1832, but his confidence was misplaced. While men and horses worked tirelessly, floundering in mud or choked with dust, and the wagons rumbled to and fro over their ways, the great bank of raw earth sank relentlessly beneath them so that their labour was as vain and as heart-breaking as an attempt to fill a leaking bucket with a teaspoon. In July 1832 Telford could venture no forecast of completion. All he could say to comfort a distracted Canal Committee, whose dwindling funds were being rapidly swallowed up in this fruitless battle, was that Shelmore was 'gradually assuming a more shapely and consolidated state'. It did not hold that shape for long. Only a month later there was a disastrous slip at Shelmore. Suddenly and with no warning a great section of the bank collapsed. It was 800 yards long and half the width of the top.

The situation was now critical. From the north end of Shelmore a continuous line of canal was now open to Ellesmere Port on the Mersey and to Manchester, for the Ellesmere & Chester Company had at last been able to complete their branch to the Trent & Mersey at Middlewich. To the south, the younger Wilsons' work on the third division was rapidly nearing completion. Work on Belvide Reservoir had not been begun because there had been delay in getting possession of the site and the Company had had to demolish and rebuild on higher ground two farms on the land which would be flooded. But to supply their top level until Belvide was completed the Company had arranged to obtain water elsewhere[1] so that this delay would not prevent the opening of the canal. Here, then, was one of the finest and most costly lines of canal ever built in England cut into two useless halves by the morass at Shelmore.

In the autumn and winter Telford was stricken down by one of his recurrent and increasingly severe attacks of sickness and the half-yearly engineering report was made by Alexander Easton. He said that Shelmore was still slipping although it had been possible to complete and puddle 300 yards of it at the southern end. The worst portion was at the meeting of the two advancing tips in the centre and here he suggested carrying the waterway in a wooden trough to enable the canal to be opened. To the Canal Committee this must have seemed a counsel of despair. As Telford was still seriously ill an urgent meeting was called at his house in February 1833 which was attended by the Marquis of Stafford, Viscount Clive, Sir James Wrottesley, Edward Monckton, James Lock and William Hazledine. In view of the state of his health, Telford agreed to the proposal that William Cubitt, a man soon to become famous as a railway builder, should deputise for him as the Company's engineer. Hazledine, Easton, the Wilsons and Provis were appointed to meet Cubitt at the Royal Victoria Hotel at Newport on the 26th of the month and to accompany him on a tour of inspection

1. This water might have been supplied from the Staffordshire & Worcestershire Canal but that Company adopted the same hostile attitude towards the newcomer as the Trent & Mersey and would not part with a drop. The Birmingham & Liverpool Committee therefore contracted in 1834 to purchase 4,000 locks of water from the Wyrley & Essington Canal Company. The consent of local millowners had to be obtained to the release of this water down the stream in the Hatherton valley (the present Hatherton Branch Canal not having been built) into the Staffordshire & Worcestershire, whence it flowed to the Birmingham & Liverpool at Autherley Junction. To obviate such water difficulties in the future it was decided in this year to double the capacity of Belvide Reservoir.

of the whole of the central division from Tyrley Locks to Church Eaton.

In his report, Cubitt expressed himself generally satisfied with the state of the works. But he recommended that the sides of the Wood-seaves cutting should be cut back further and that hard core should be tipped at Shelmore. He declared with all the confidence of youth that the battle with the great bank would be easily won in this way. The hard material would, he maintained, sink down through the marl and sand and soon form a new firm base which would not yield under the weight. The Committee were by now reduced to that state of mind which clutches at any straw and Cubitt's confidence put fresh heart into them. In May, after another meeting at Telford's house, the edict went forth that 40,000 cubic yards per month were to be laid on Shelmore Bank for as long as Cubitt should think fit.

In July 1833 the Wilsons finished their work on the third division with the exception of Belvide Reservoir, where their whole labour force was now at work on the dam head. The Newport branch was finished also and on the 17th of the month the first boat went down the new canal and on to the Shrewsbury Canal at Wappenshall. But still the fruitless and apparently endless filling went on at Shelmore. It was necessary to extend the railway and buy land from 'the Clergyman at Norbury' in order to obtain additional supplies of suitable material. The bank, said Cubitt, was up to full height except for a distance of 400 yards in the middle and should be completed in three months. Yet in October he could only say: 'it certainly takes more earth to make good than even I had anticipated'. He hazarded no forecast of completion and it is doubtful whether the Committee was comforted by his vague assurance that there was 'nothing to alarm'.

On 16 January 1834, Cubitt reported that the puddled bed for the canal on Shelmore Bank was almost complete, but he added the ominous note that there were 'slight' signs of shifting for fifty or sixty yards by Shelmore Farm. This was at the same place of greatest danger where the two tips met and where, therefore, it had been most difficult to consolidate the ground as the banks rose. At the end of March, after an interval of two years, Telford rallied his strength sufficiently to make the last of so many journeys from London to Shrewsbury. There Cubitt met him and together they travelled to Shelmore where the old man, frail now and very deaf, looked for the last time at the great bank which had, as he expressed it, 'caused so much trouble, expense and procrastination'. He thought of Lord Anson's precious pheasants, counted their cost and wrote bitterly: 'Had the original line laid down in the Parliamentary Plans been adhered to, the canal would not have exceeded the estimate.'

Sadly he turned his back on the scene of his disappointment and returned to London. It was up to young Cubitt now, and Easton and Provis; he had played his last and hardest innings and for him the game was nearly over. No doubt he would live to see his work finished for it only remained to close the gap in Shelmore Bank. But at the end of May Shelmore slipped again, and as if this were not enough, a mile to the north 10,000 cubic yards of marl and rock came thundering down from the side of Grub Street cutting near Blakemere Pool and obliterated all trace of the canal for a distance of sixty yards. Provis's men fought back stubbornly; the cutting was cleared; the sides cut back; the banks of Shelmore rose yet again despite still further slips on both sides. But Shelmore Great Bank had successfully defied Telford and foiled his hopes. On 2 September 1834 he died with his canal still divided into two useless halves. Shelmore had even forced Cubitt to change his confident tune, but at least he was honest. 'I am exceedingly mortified', he wrote in his October report, 'to think I can speak with so little certainty as to any fixed time for completion.'

In January 1835 he was more confident and on Monday 2 March, exactly six months after Telford's death, the first boat was able to pass through the full thirty-nine miles of his Birmingham & Liverpool Junction Canal. It left Autherley Junction at 8 a.m. and John Freeth, the Birmingham Canal Secretary, was one of the party of officials on board. He reported that the state of the Shelmore embankment was 'still very precarious', particularly on the east side and that it was only with difficulty that a fully loaded boat could have been got over it. The channel was still only a boat's width in places. But the long battle had evidently been won at last, for in the following July Cubitt was able to report that the canal was now at full width over Shelmore, that the condition of the bank was steadily improving and that grass and rushes had been planted to consolidate the slopes and break the wash of the boats.

As the grass and the rushes grew up to seal and heal the raw earth of Shelmore Bank, so the recollections of the bitter six-year struggle to build it faded until they vanished out of memory. The boatmen who swing their boats round that long sweeping curve today and look out from its height across the plain to the steep cone of the Wrekin know nothing of its history. But on William Cubitt and his fellow railway engineers the lesson of Shelmore was not lost.

EVENING IN ABINGDON STREET

IT was not until 1821, when he was sixty-four, that Telford acquired a home of his own. For twenty-one years the Salopian Coffee House at Charing Cross was his London headquarters and it became a regular meeting-place for his assistants and for fellow-engineers visiting London from the provinces or from abroad. Several rooms were permanently reserved for his use and he could readily command extra accommodation whenever he wished to entertain. It was doubtless an excellent arrangement for a man who was seldom in London for more than a month or two at one time. Here Davidson's three sons foregathered whenever they were in London, and indeed any of Telford's acquaintances and fellow-countrymen were always sure of a welcome at the Salopian. James Davidson, who followed his father's profession, lived there with Telford for some time. Telford always took a fatherly interest in his friend's three boys and on one occasion Davidson wrote to his eldest son: 'you should not neglect to write to Mr. Telford letting him know what you have been doing . . . asking his advice &c; if he do not notice you – never mind – keep writing away – there's nothing the matter, only getting old and cross like your grey headed Dad'. The young poet, Thomas Campbell, also shared Telford's hospitality at the Salopian for some time. 'Telford is a most useful cicerone in London,' wrote Campbell to a friend in 1802, 'He is so universally acquainted, and so popular in his manners, that he can introduce one to all kinds of novelty, and all descriptions of interesting society.' In a letter to Andrew Little, Telford expressed a glowing opinion of Campbell's talent and forecast a great future which the young man failed to fulfil. Obviously Telford was better at scenting engineering than poetic ability. According to Smiles, Campbell wrote *Hohenlinden* while he was staying at the Salopian and the two men collaborated in polishing the poem. If the result did not achieve greatness, at least it was to become required learning for generations of children.

Telford brought a great deal of business to the Salopian and successive landlords looked upon him as part of their property. He was actually sold as the most important item in the goodwill. It is said that when he finally announced his departure, the unfortunate landlord, who had only recently taken possession, exclaimed, 'What! leave the house! Why, Sir, I have just paid £750 for you!' But at sixty-four the desire for a place of his own was too strong and so Telford moved to No. 24 Abingdon Street, a house opposite the Houses of Parliament which is at present occupied by the Lands Commission. Having narrowly escaped destruction by bombing, it may soon be swept away by the Westminster improvement scheme.

Telford took an almost childlike pleasure in this, his first house, and he could scarcely have found a home more suitable to his needs or more appropriate in its associations. It was conveniently near the House when any Parliamentary business was afoot and it had previously housed two other illustrious men, Labelye, the Swiss architect of Westminster Bridge, and Sir William Chambers whom Telford had served as a stonemason so long ago. In Telford's day a painting of Westminster Bridge by Canaletto, empanelled above the chimney-piece in his dining-room, survived as a memorial to Labelye's occupancy. The house soon became a school for civil engineers. Rooms were set aside for Telford's younger assistants and pupils, and there they lived and worked. Among them, though he did not become an engineer, was James Little, son of Andrew's brother William, and Joseph Mitchell, the son of John, Telford's 'Tartar' of the Highland roads. Both young men have left us glimpses of life at Abingdon Street.

'We breakfast at 8', Little told his mother in 1825, 'and dine at ½ past 5 – this would not suit you, but now that I am accustomed to it, I like it very well. Mr. T. is very fond of the mutton ham which is only presented on great occasions, & Mr. Anderson (another young assistant) & I get a little now and then by way of a treat – we have large dinner parties sometimes, & a fine *set out* there is – but do not you imagine that Mr. T. eats nothing but cake, & drinks nothing but water, if you do, you are quite mistaken; we drink wine *every day*. . . .'

Joseph Mitchell recalled his Abingdon Street days in his reminiscences: 'Our working hours were from nine till seven; the evenings we could dispose of as we liked. The old gentleman treated us as sons.

If he had friends we dined in their company. His house was handsome ... and he was very proud of it. After having lived in lodgings and hotels all his life, he appeared gratified and delighted at the additional comfort and convenience of his new residence. When any friend called and complimented him on his new quarters he used to say "Oh, you must see my house", taking them over the principal apartments, pointing out the solid mahogany doors and marble chimney pieces. "Below", he would say "I have a perfect village"; and opening the door of the apartment in which we worked, he added "and here are two raw Scotchmen"....

'Telford did not go into what is called "Society",' Mitchell continues, 'but he was always delighted to see his friends at home. Occasionally he had dinner parties consisting of gentlemen chiefly connected with his works in progress, such as Sir Henry Parnell, afterwards Lord Congleton, Milne, Secretary, afterwards Chief Commissioner, of the Board of Works, Admiral Sir Pulteney Malcolm, and General Pasley. Southey and Campbell the poets were also frequent guests, and he had visits from agents and engineers in the country. Telford was the soul of cheerfulness, and used to keep his guests in a roar of laughter. He had a joke for every little circumstance and he was full of anecdote...'

Shortly before Telford's move to Abingdon Street, some of the younger engineers began to voice the need for some established organisation which would be of more practical value to their profession than the exclusive and almost purely social Society of Civil Engineers which was at this time dominated by the Rennies. The driving force behind this movement was Telford's young assistant H. R. Palmer whose name has been mentioned earlier in connection with the railway and canal traction experiments. Henry R. Palmer had been apprenticed to Bryan Donkin, subsequently became resident engineer to the London Docks, and has been described as a man of great inventive powers who would have achieved greater fame had he not died prematurely at the age of forty-nine. When he joined Telford in 1818 he was only twenty-three and it was on 2 January in this year that he invited several other young engineers to meet him at the King's Head Tavern in Cheapside. They were William Maudslay, Joshua Field, James Ashwell, Charles Collinge and James Jones. The six men hammered out a set of rules and the new society was formally established. It is easy to start a society but, as Palmer found, it is much more difficult to make it grow. For the next two years the little band of founder members continued to meet in taverns or coffee houses, but it must have seemed that the society had been still-born, for in all this time only four new members were attracted, one of them being William Provis. But at a meeting held on 25 January 1820 Provis proposed that Telford should be invited to become the first President of their Institution of Civil Engineers. The motion was carried and the invitation dispatched. Telford, who, it is said, knew nothing of Palmer's evening activities, pondered the invitation for some time but finally signified his agreement in March. His acceptance set the seal of success on Palmer's efforts.

The Institution no longer met at Gilham's or the Kendal Coffee Houses, but acquired a home of its own, first in Buckingham Street, Strand, and later in Cannon Row, Westminster. Telford was no figurehead President. Progress was still neither easy nor rapid and many engineers held aloof from the new Institution even if they were not actively opposed to it. Had it not been for Telford's efforts and the weight of his influence the new body would almost certainly have foundered. He was determined that it should succeed and become neither a social mutual admiration society nor a market-place for engineering talent, but a common pool of knowledge and experience which would promote the advancement and recognition of the new profession.

As soon as the Institution acquired its own premises in Buckingham Street a series of weekly meetings began at which a member would read a paper on some engineering subject. It became Telford's custom to hold a dinner party at Abingdon Street every Tuesday evening and then take his guests along to the meetings. In this way he was able to introduce influential men, many of whom joined the Institution as ordinary or honorary members. Young Joseph Mitchell was always a member of these Tuesday parties, for Telford had assigned to him the task of taking notes of the discussion which followed the reading of a paper and in this way the Institution's *Proceedings* began. Similarly, it was Telford's presentation of books and drawings from his collection which laid the foundation of the Institution's splendid library. His enthusiasm for the Institution is revealed in his letters to Count von Platen. He had not been a year in office when he was asking for drawings and documents from Sweden for the library. 'You will 'remind them', he writes, 'to send me drawings and descriptions of *works really executed.* We have no wish for learned discussions; *Facts* and *practical operations* are to compose our collections and we should leave *project* and *theory* to those who are disposed to create new systems.'

Exactly a year later, in March 1822, he thanks von Platen for 'the drawings and communications for the Engineer Institute' and then gives him the following progress report: 'There are now above 50 members in England, Scotland, Ireland, Holland, France, Germany, Sweden and the East Indies. With these much may be accomplished if the Members can be rendered effective. Perseverance will in time render the collection of Books, Manuscripts and Drawings invaluable. This end accomplished, the rising generation will, for their instruction and benefit, be attracted to this establishment as a centre. We admit only what has really been performed – no theoretic projects. By the by, we shall circulate a Catalogue of the books and documents in the Institution Rooms so that each member may judge of the value of the collection ...' 'The Civil Engineer Institute is going on very prosperously', he reports in July 1824, and in his final letter to the Count dated 22 January 1827 he refers to it as 'now numerous and respectable' – adding the gentle reminder that Mr Edstrom had not paid up his annual subscription! This last letter of Telford's concludes with the news that: 'I was last evening admitted a Fellow of the Royal Society of London.' This distinction still further increased the prestige and the influence of the Institution's President and thus enhanced the standing of the Institution itself. In the following year it was awarded its Royal Charter which signified the national recognition of civil engineering as a profession of honour and repute. One of the Institution's rules for which Telford was responsible stipulated that each member should submit a paper on some practical engineering development once a year. This was never in fact enforced, which was just as well in view of the Institution's present number of members, but the idea was sound enough in these earliest days.

After 1830, when the Institution was firmly established, Telford played a somewhat less active part in its affairs. As well as recurrent bouts of illness, increasing deafness now troubled him. It cut him off from his fellows and, as Joseph Mitchell recorded, made him self-conscious in their company. No other infirmity can so exacerbate the loneliness of old age. Without ears communication fails and so there was an end to the dining and wining at Abingdon Street. He could no longer set the table in a roar with his sallies but could only gaze blankly from face to face as conversation flickered inaudibly to and fro. The mind behind the barrier was as keen as ever, but the furious tempo of Telford's life perforce slowed down and the old house grew quiet. Finding himself forced to dwell more and more upon his memories yet still needing some occupation for his restless mind he resolved to write his autobiography, and this work filled his days to an increasing extent.

As a personal record the result of this last effort is valueless. Telford's brief, factual descriptions of his works coupled with the magnificent illustrations and drawings in the accompanying atlas no doubt fulfil his intention of leaving a record which might be useful to future engineers, but they tell us nothing of the man himself, so that the book cannot truly be called an autobiography at all.

Even if deafness had not isolated him Telford must have been doomed to loneliness in old age, that inevitable penalty which a man must suffer who dedicates his life exclusively to his work. For although he had many friends he had no intimates. With the possible exception of Matthew Davidson, Andrew Little appears to have been his only confidant. After Andrew's death Telford continued to correspond with William Little and his family. He would ask for news of Eskdale, send money to be distributed anonymously by the Littles to deserving kinsfolk, and help the people of his native valley in numerous other ways. But he never reveals himself in these later letters as he once did to Andrew. We know from these letters to Langholm that he still kept green the memories of his childhood by the Megget Water; we are told that he continued to read widely, not only in English but also in French and German; we know from his friendship with Southey and Campbell that he had not lost his love of poetry, but beyond these bare facts we know nothing. Telford is silent. Here, as inscrutable as his marble statue in Westminster Abbey with its pupilless gaze, stands the greatest engineer England had ever known; Knight of the Royal Order of Vasa, Fellow of the Royal Societies of Edinburgh and London, First President of the Institution of Civil Engineers. But what of the old man beneath this impressive mantle of greatness? What were his private thoughts? As he looked back over the years did he feel that the game he had played so hard had been worth while?

Just three months before his death, a pencil drawing was made of Telford by William Brockedon. The thick curly hair has gone and the craggy face has softened, but it still tells us nothing; the dark eyes do not look 'at us; indeed, like those of a man in reverie they are not focused upon any external thing, nor do they betray any secret of the mind within. Notwithstanding the tributes of those who knew him, the portrait of this solitary, enigmatic and proud old man which emerges from print, pencil and pigment is not a particularly lovable one; it is too lacking in warmth and sympathy – even in evidence of common human frailty – to claim our affection. Perhaps it is due to egoism that while the engineer's greatness appeals to the intellect the man fails to move the heart; perhaps it is simply because the private man so sedulously evades our question. Only by guesswork can we etch pupils on the eyes of that marble face in Westminster and make them expressive of the man within.

Telford's two previous biographers, confronting a similar problem, have painted a picture of a long life drawing to a tranquil close. The great engineer, a benevolent old man loaded with honours and with many friends about him, sinks to his rest happy in the recollection of a long and successful career in which he has seen all his great works completed. It is a pretty picture but, like the traditional happy ending of the fairy story, it is a little too good to be true. Life is not like that. For one thing we know now that it is not correct to say that he lived to see all his major works completed. The fact that Shelmore Great Bank cheated him of this consummation and taunted him with failure up to the hour of his death has been either forgotten or ignored. Yet it is surely much more true to life. No matter what we may say to the contrary, life is often mysteriously just. For every fault there comes a reckoning. Though he was never arrogant, Telford was at heart a very proud man and the prideful must always pay a heavy price sooner or later. In his case that account was so long deferred that it must have seemed that it would never be rendered. Yet it came in the end. Shelmore Bank was a part of the payment. His treatment by the Liverpool & Manchester Railway Company and the outcome of the Clifton Bridge episode represented two other equally bitter instalments. Although Telford ultimately defeated the Liverpool & Manchester directors, their offhand treatment of his assistants and the contemptuous remarks made about him in their widely circulated statement must have cut very deep. It was behaviour unprecedented in all his long life and it must have been to him an acutely painful reminder that a new generation of engineers had come to the fore who had little or no respect for his long experience and with whom he was no longer *persona grata*. The lesson of the Clifton Bridge competition was the same. The younger Brunel, a mere stripling of twenty-four, had treated him with equally scant respect; had not scrupled to inflict a most humiliating defeat upon an old man who had so lately been acclaimed the greatest bridge builder the world had ever known. These things must have impressed upon him the bitter truth that there comes a time when the strongest and surest hand must falter and that for him the moment had come when he must take his last bow and leave the stage for ever. He had lived too long.

So the old engineer went down before the triumphal onslaught of the railway engineers. But not without honour. He had directed that he should be laid without ceremony in the burial ground of his parish church of St. Margaret, Westminster, but at the representation of his fellow members of the Institution of Civil Engineers the nave of Westminster Abbey was his resting place. 'I hold', he had said once, 'that the aim and end of all ought not to be a mere bag of money, but something far higher and far better' and he acted upon this precept all his life. It was the challenge of the task which appealed to him, not the reward and often, as in his work on Highland harbours for the Fisheries Society, he accepted no fee at all. Consequently his estate of £16,600 was small considering his long lifetime of intense activity. Two thousand pounds and all his books, drawings and papers went to the Institution of which he was so justly proud; a similar sum he bequeathed in trust to the parishes of Langholm and Westerkirk for the establishment of libraries so that his kinsfolk in the dale should never be denied the books he loved. The rest of his fortune he divided among his friends, his young assistants and surveyors. Among them, Robert Southey, in dire financial straits at the time, blessed the memory of his friend for the unexpected legacy of £850.

When, so soon after his death, England went railway mad, Telford and his works were eclipsed. The proud canal companies he had served were humbled, were forced to beg for traffic and ultimately to sell themselves to the all-conquering railways. Dazzled by wonders such as the Britannia and Saltash Bridges the world forgot Pont Cysyllte and the Menai. The great road through Nant Ffrancon fell silent and Ogwen echoed no longer the cry of the post-horn. Yet Telford's faith in his roads would be justified by time and they would take their revenge. Even in the Institution which he had helped to create attempts were made to diminish Telford's stature. Notwithstanding their collaboration with Telford in the Fens, the Rennie brothers perpetuated their father's hostility towards him. But the time soon came when they could not afford any longer to ignore the Institution. Both joined after Telford's death and Sir John eventually became President. He took the opportunity in his Presidential address subtly to disparage Telford and his work, claiming amongst other things that it was John Smeaton who was entitled to be called the Father of Civil Engineering. That a dead man should thus have been made the instrument of Sir John Rennie's animus is saddening, and it in no way diminishes John Smeaton's stature as an engineering pioneer to hold that Rennie was wrong. This book should be proof enough that it was Thomas Telford who, not only by his work for the Institution but by the example of his whole life and the genius of his organisation and command of great enterprises, has the right to that honour. A quarter of a century after Telford's death that great railway engineer Robert Stephenson gave the lie to Rennie and paid an eloquent tribute to his predecessor when he directed that his body should be laid beside that of Telford in the nave of Westminster.

There, side by side, lie the great road makers. The question still stands: was it all worth while? Perhaps we have the answer in some lines which Telford added, at some unknown date, to the end of his poem in praise of Eskdale. He wrote:

> Yet still one voice while fond remembrance stays
> One feeble voice shall celebrate thy praise,
> Shall tell thy sons that, whereso'er they roam,
> The hermit Peace hath built her cell at home,
> Tell them, Ambition's wreath and Fortune's gain
> But ill supply the pleasures of the plain;
> Teach their young hearts thy simple charms to prize:
> To love their native hills and bless their native skies.

A NOTE ON SOURCES AND ACKNOWLEDGEMENTS

THE three chief printed sources on Telford's life to which I have referred are, in chronological order, the *Life* written by himself, edited by John Rickman and published in 1838; the section devoted to Telford in Volume 2 of the *Lives of the Engineers* by Samuel Smiles and *The Story of Telford* by Sir Alexander Gibb, published in 1935. Telford's autobiography is a useful source of factual information concerning his works, but as a personal document it is valueless. It is companioned, however, by an *Atlas* fully illustrating Telford's works which is a truly magnificent production. Smiles's account of Telford's life is brief and readable but, like all his work, it is written in that uniformly adulatory style reminiscent of a funeral oration or an eighteenth-century epitaph which is no true revelation of personality. We know that Telford and Rennie were at odds with each other, yet Smiles makes both men appear such super-human paragons of virtue that no such clash of personalities is conceivable. In Sir Alexander Gibb's biography, Telford's life is treated in strict chronological sequence. As the engineer's activities were so manifold, and as many of his important works were carried on simultaneously over a long period of years the effect of this treatment is to make the book difficult and at times bewildering reading. Nevertheless it is the product of most painstaking research and is therefore of the greatest value as the only complete record of Telford's work. I am very deeply indebted to its author.

When Telford's own *Life* was published after his death, his friend Robert Southey wrote a very long review of it in the *Quarterly Review* but most of the information contained therein is taken from the book so that it contributes practically nothing to our knowledge. Of greater value to the biographer is Southey's *Diary* of his tour with Telford in the Highlands. The original Diary is in the Library of the Institution of Civil Engineers, but I have made use of the printed version published by John Murray in 1929.

Through the courtesy of Sir Alexander Gibb I have had access to a volume containing copy letters from Telford to Matthew Davidson, from William Jessop to Telford and from Matthew Davidson to his sons. All quotations I have used from such letters come from this source.

I am deeply indebted to Mr James Little of Craig, Langholm, who most generously lent me volumes containing copies of all the letters written by Telford to Andrew Little and other members of the Little family. These are of great value as the only truly personal record which Telford has left to us and I have drawn upon them very freely. Unfortunately they only illuminate the earlier period of his life. It would appear that, as in the case of I. K. Brunel, when Telford's career was at the flood he no longer had the time to commit his private thoughts to paper. It is when a man is seeking fame that he confides his ambitions; when he achieves it he falls silent.

When Sir Alexander Gibb wrote his biography he had access only to the letters written by the Count von Platen to Telford which are among the Telford papers in the Library of the Institution of Civil Engineers. Recently, however, Mr Eric de Maré discovered the letters written by Telford to von Platen in the archives of the Gotha Canal Company at Motala. The copies of these which Mr de Maré has presented to the Institution's Library not only illuminate the part played by Telford on the Gotha Canal but provide fresh first-hand information concerning his other activities during this long period.

Among other Telford papers in the Institution's Library which I have studied are dossiers of letters and reports on the following subjects:

Highland Surveys.

Holyhead Road Survey and Menai Bridge.

South Wales Road Surveys.

Correspondence and reports on Stratford & Moreton, Stockton & Darlington, Newcastle & Carlisle and Liverpool & Manchester Railways.

Trent & Mersey Canal: Harecastle New Tunnel and Knipersley Reservoir.

Report on the opening of Pont Cysyllte Aqueduct.

My accounts of these works are based on these documents and the relevant quotations come from them. The Trent & Mersey dossier is particularly valuable because the Minute Books of the Trent & Mersey Canal Company are missing.

Other original items in the Library which I was able to study were Telford's *Treatise on Mills*, his Architectural Notebooks and his Portfolios of original drawings. The latter include his original designs for the Runcorn and Clifton suspension bridges. I am deeply grateful to the Council of the Institution of Civil Engineers for their permission to carry out this research in their Library and for the kind assistance of the Library staff.

Unless otherwise stated in the text, my account of the progress of the works of the Caledonian Canal is based upon information obtained from Volumes I and II of the Reports of the Commissioners of the Caledonian Canal. These incorporate Telford's own reports which he issued biannually in spring and autumn until May 1820, after which James Davidson became responsible. For enabling me to study these volumes in the offices of the Caledonian Canal at

Clachnaharry, Inverness, I am indebted to the British Transport Commission and to Mr F. Whyte, O.B.E., M.C., the Canal Manager and Engineer.

Since the last biography of Telford was written, research into the history of Telford's canal works has been made infinitely easier thanks to the collection by the British Transport Commission of the relevant Company Minute Books and other documents under one roof. I am grateful to the Commission's archivist, Mr L. C. Johnson, and his staff for enabling me to study the following:

Minute books of the General Assembly and Committee of Management of the Company of Proprietors of the Ellesmere Canal, the Birmingham & Liverpool Junction Canal and the Macclesfield Canal.

Letters from Telford to John Freeth relating to the Birmingham Canal improvement scheme.

My accounts of the building of the Ellesmere and the Birmingham & Liverpool Canals have been almost exclusively based upon information obtained from these Minute Books. In the case of the Macclesfield Canal, the first volume of the Minutes of the Committee of Management covering the period of construction is unfortunately lost, but I obtained sufficient evidence to satisfy me that the part played by Telford was confined to the preliminary survey.

For the loan of reports by William Cubitt on the Birmingham & Liverpool Junction Canal and of the Specification of Messrs Rowland & Pickering's Patent Canal Lift, I am indebted to Mr Charles Hadfield, and to Mr Eric de Maré for much help and advice in connection with my chapter on the Gotha Canal with which, alone of Telford's canal works, I am not personally familiar.

The letters from Telford to the firm of Boulton & Watt, including Telford's personal letter to James Watt concerning John Rennie, are in the possession of the City of Birmingham Reference Library where I was able to consult them.

In conclusion I would like to express my gratitude to the following for their help and advice: Messrs. J. Emlyn Jones, O.B.E., J. D. W. Jeffery and A. D. Holland of the Ministry of Transport and Civil Aviation for information concerning Telford's road bridges; also to Messrs. Edward Armstrong of Langholm, D. S. Barrie, John Betjeman, Christopher Hussey, M. M. Rix,* Michael Rogers, J. T. Smellie, Graham Webster and E. A. Wilson.   L.T.C.R.

* Mrs S. M. Rolt gratefully acknowledges additional material incorporated on pages 32, 33.

BIBLIOGRAPHY

THE following is a list of works consulted, excluding the original sources listed in the bibliographical notes:

COLVIN, H. M., *A Biographical Dictionary of English Architects, 1660–1840*. London, John Murray, 1954.

DE MARÉ, ERIC, *Swedish Cross Cut (The Gotha Canal)*. Sweden, A-B Allhem, Malmo, 1957.

GIBB, SIR ALEXANDER, *The Story of Telford*. London, Alexander MacLehose, 1935.

HADFIELD, E. C. R., *British Canals*. London, Phoenix House, 1950.

HADFIELD, E. C. R., *Canals of Southern England*. London, Phoenix House, 1955.

MARSHALL, C. F. DENDY, *A History of British Railways Down to the Year 1830*. London, Oxford Univ. Press, 1938.

PRATT, E. A., *Scottish Canals and Waterways*. London, Selwyn & Blount, 1922.

PRIESTLEY, JOSEPH, *Historical Account of the Navigable Rivers, Canals and Railways Throughout Great Britain*. London, Longman, Rees, Orme, Brown & Green, 1831.

SMILES, SAMUEL, *Lives of the Engineers*, Vol. II. London, John Murray, 1862.

SOUTHEY, ROBERT, *Journal of a Tour in Scotland in 1819*. London, John Murray, 1929.

TELFORD, THOMAS, *Life of Thomas Telford*, ed. John Rickman. London, James & Luke G. Hansard, 1838.

INDEX

Aberdeen: harbour improvements, 89
  Don bridge, 148 n.
Ackerley, Samuel, 59
Acrefair Colliery, 68
Adam, Robert, 25, 26, 35, 39
Adderley Leys, 189, 191
Adelaide Gallery, 171
Agriculture, Telford on, 76
Akers, mail-coach superintendent,
  140
Albion Mills, 28
Aldersley Junction, 180
Aldrich, Dr. Henry, 42
Alison, Rev. Archibald, 27
Alyn, River, 68
Anglesey: poor roads in, 124; Telford's
  plan for new roads, 125
Anglesey, First Marquis of, 133
Anson, Lord, of Norbury Park, 190,
  196
Anstruther, Colonel, 79
Archaeology, eighteenth-century, 36–7
Architecture, eighteenth-century, 34–5
Argyll, Duke of, 78
Arisaig, 82
Arran, Isle of, 101
Ashwell, James, 200
Ashworth, Yorkshire worker on
  Gotha Canal, 114, 116
Atherton, Charles, and Dean Bridge,
  150
Audlem, 190; locks, 189
Autherley, 189
Avonmouth, 76

Badenoch, 78
Bagge, Samuel, 108, 112, 113, 114;
  death, 115

Baird, builder of Edinburgh and
  Glasgow Union Canal, 165
Bala, Lake, 71
Baldwin, Thomas, 41
Ballater Bridge, 86
Baltic timber, Government duties on,
  105
Banavie, 101; lock-building at, 101
Banbury, 50
Banff, harbour improvements to, 89
Bangor, 124
Bangor (County Down), harbour
  improvements, 144
Barlow, Peter, 131, 171
Barnes, surveyor of Grand Union
  Canal, 164 n.
Barnet, road improvements, 126
Barrow-in-Furness, 176
Barry, Wolfe, 138
Bath: Telford's work at, 40–41; his
  criticisms on architecture of, 40–41;
  failure of City Bank, 41; Lansdown
  Crescent, 41; St. James's Square,
  41; Theatre Royal, 41
Bath, Lady, 58
Beachley, suggested suspension bridge
  at, 146
Beachley-Aust; improvement to Old
  Passage Ferry, 146–7
Beaufoy, Henry, 78, 123
Beauly Firth, 94
Beauly, River, 79; bridge over, 83
Beauly sea lock, 99
Beckford, William, 42 n.
Beeston Castle, 70
Bell, Henry, 105
Belvide Reservoir, 189, 194 n., 195
Bentham: State of Architecture, 35

Berg, 111

Berkeley Pill, 157, 158

Bernadotte, Marshal, 111–12

Bersham: Colliery, 68; Ironworks, 101

Berwyn-Glyn Dyfrdwy Road, improvements, 128

Berwyns, the, 124

Bettws-y-Coed, 124; Bridge, 128

Bevan, Benjamin, 158, 164 n., 171

Bewdley Bridge, 47–8, 149, 160

Birkwood Burn Bridge, 89

Birmingham: Telford's criticism of, 42–3

Birmingham Canal, 179–81

Birmingham Canal Company, 179, 180, 182

Birmingham and Fazeley Canal Company, 50

Birmingham and Liverpool Junction Canal, 163; Nantwich Basin-Turley, 188–9, 195; Turley-Church Eaton Bridge, 189, 190; Church Eaton Bridge-Autherley, 189, 192; Grub Street cutting, 190, 191; Shelmore Bank, 190, 192–7; trouble on Nantwich Bank, 191; and on Edleston Bank, 191; first boat on, 196

Birmingham and Liverpool Junction Canal Company, 183; opposition from landowners, 188; compensation paid by, 188

Birmingham-Wednesbury, canal section, 180–81

Blackfriars Bridge (Mylne's), 157

Blair Atholl, 78

Blake, William, 44

Bloomfield, 181

Bonar Bridge, 85, 86 and n., 87, 88, 160

Boren, Lake, 111

Borrow, George, 60

Bosley, 179

Boston Bridge, 160

Bosworth tunnel, 175

Bouch, Thomas, 87

Boulton, Matthew, 28

Boulton and Watt, 28, 92; beam pumping engines, 92, 98, 100, 102, 155, 175

Braemar, 78

Bramah, Joseph, 170

Bramah hydraulic press, 131

Braunston, 182; road improvements at, 126

Brickhill, road improvements at, 126

Bridge-building, 47, 82–3; in Highlands, 86 ff.; suspension bridges, 131; and storms, 151–2

Bridgewater Canal, 50

Bridgewater Canal Company, 164

Bridgnorth: Church of St. Mary Magdalene, 38–40, 58; ironworks, 66; Bridge, 47–8

Brindley, James, 34, 50, 59, 132, 164, 173, 180; his canal system, 51–2; as canal designer, 184–5

Brinklow, 182

Bristol, Telford on, 76

Bristol–Milford mail route, 146

Bristol–Uphill Bay, suggested new road, 146

British Fisheries Society, 77, 80

Brockedon, William, 204

Broomielaw Bridge, 150–51, 160, 163

Broseley, 118

Brown, Captain (later Sir Samuel), 132

Brown, Thomas, of Halesowen, 168

Brown, Thomas, of Manchester, 179

Brunel, Isambard Kingdom, 26, 49, 76, 181, 205; and Clifton Suspension Bridge, 152

Brunel, Sir Marc, 131

Brunton, William, Patent Chain Cable Works, 131

Brymbo Colliery, 68

Buccleuch, Third Duke of, 20

Buck, G. H., 171–2

Bude Canal, 156–7

Buildwas Bridge, 47, 48–9

Bulkeley, Viscount Warren, 133 n.

Butterley Ironworks, 99, 102

Cadnant Brook, 136

Caernarvon: opposition to Menai Bridge in, 133; day coach, 140

Caithness, 82, 85

Caldwell (Trent and Mersey Canal Committee Chairman), 166–7, 174, 185, 187

Caledonia sloops, 99, 100

Caledonian Canal, 92 ff.; reasons for building, 94–5; work on eastern and western terminals, 95; difficulties with labour force, 96; opening, 104; cost, 104; time taken, 104; commercial failure, 105; criticism of, by public and parliament, 105; engineering defects, 105–6; restoration of, 106; use of, in World War I, 106

Caledonian Canal Commission, 82–3, 95

Cambridge: Pembroke College, 41n.

Campbell, Colin: *Vitruvius Brittanicus*, 35

Campbell, Thomas, 198

Canal-building, 34, 50 ff.; use of inclined plane, 52; aqueducts, 59–60; end of popularity of, 67; problem of summit level supplies, 68; vertical canal lift, 68–9; revival after 1824, 165

Canals, working conditions on, eighteenth century, 171

Capel Curig, 125

Cardiff–Cowbridge road, 146

Cargill, John, 87, 149; and work on Caledonian Canal, 96, 101–2

Carl XIII, King of Sweden, 111

Carlisle, Thomas, of Fort William, 114

Carlisle–Glasgow Road, 89

Carlisle–Port Patrick, suggested road, 89

Carmarthen–Milford mail-coach road, 145

Carracci, Annibal, 42

Carreghom, 58, 59

Carron, Loch, 85

Cartland Crags Bridge, 90

Cauldkine Rig, 17

Cefn Mawr, 72

Ceiriog, River, 54, 55, 59, 124; aqueduct over, 65–6

Ceirw, River, 124, 128

Chambers, Sir William, 25, 35, 39, 114, 199

Chapman, William, 171

Cheltenham petition on communications with Ireland, 145–6

Chemistry, as applied to art of building, 35

Chester, 53, 59; Grosvenor Bridge, 178

Chester Canal, 53, 70

Chester Canal Company, 53, 164

Chester–Bangor Road, survey of, 126

Chirk: aqueduct, 65–6; tunnel, 65, 177

Chirk–Holyhead road survey, 126

Clachnaharry, 95; lock-building at, 98–9

Clifton Bridge, 152–3, 204, 205

Clive, Viscount, 194

Clowes, William, 60, 61

Coalbrookdale: Bridge, 49; Foundry, 48, 52

Coalport, 52

Cobbett, William, 76

Coleham Foundry, 66–7

Coleham link test, 135

Coleridge, S. T., 90

Collinge, Charles, 200

Colvin, H. M.: *Biographical Dictionary of English Architects*, 42 n.

Congleton, 179

Conon, River, 79; Bridge, 86 and n.

Conway, River, 128; Bridge, 138–9; opening of, 141; valley, 124–5

Corpach, 95; brewery built at, 96; lock-building at, 100; demand for sea transport, 100

Corpach sloop, 100

Corrieyarrach, Pass of, 78

Cortney, William, and Company, 100

Corwen, 124

Coseley tunnel, 181

Cosford, road improvements at, 126

Coventry Canal Company, 50

Cowley tunnel, 190, 192
Cowper, William, 30
Craig Phadrick, 99
Craigellachie Bridge, 86, 87, 138, 160
Crick tunnel, 175
Crief, 89
Crinan Canal improvement scheme, 88
Cromarty, 80
Cromford and High Peak Railway, 69 n.
Crowland, 162
Cubitt, William, 194–7
Cullen, harbour improvements to, 89
Culloden Moor, battle of, 77
Cumberland, 'Butcher', 77
Cupar-in-Angus, 78
Currie, Dr. James, 80

Dalziel brothers, the, of Edinburgh, 18
Dance, Sir Charles, 170
Dance, George, the younger, 41, 153
Darby, Abraham, 48, 52, 55
Darwin, Erasmus: *The Botanic Garden*, 44, 45
Davidson, James, 98, 198
Davidson, John, 97, 98
Davidson, Matthew, 24, 40, 47, 58, 62, 63, 64, 65, 73, 74, 87, 134, 203; work on the Caledonian Canal, 95, 99–100; character, 96–7; contempt for Highlanders, 97; death 100, 188
Davidson, Thomas, 97, 98, 100
Davies, David, 140
Davies, Hugh, 138
Davies, William, 65
Ddu, Robin, of Bangor, 129, 141
Deadman, Samuel, of Inverness, 99
Dean Bridge, 150
Dee, River, 53, 59; bridge over, 83; Valley, 124
Deepfield, 181
Dempster, George, 78, 123
Denson, Thomas, 72
Denver, 161
De Ville, James, 140
Diana, the, 110, 112
Dingwall, 85

Dixon, Colonel, 81
Donaghadee harbour improvements, 144
Dochfour, 102
Dochfour, Loch, 99
Docks, work on, 76
Doddington, 28
Donkin, Bryan, 102, 118–19, 131, 170, 171, 200
Donnington Wood Canal, 52
Dornoch, 88
Dornoch Firth, 85, 86
Doune, 89
Dounreay Atomic Plant, 86 n.
Downie, Murdoch, 94
Drainage Commissioners, 161
Dredgers, steam bucket, 102, 118
Duich, Loch, 94
Dunkeld, 79; Bridge, 86, 87
Dulsie, 79
Duncombe, John, 53, 55, 56, 58, 70, 84; Telford's criticism of, 84
Dundee, harbour improvements, 89
Dutton and Buckley, 177

Eastburn, Mr. of Odiham, 54
East India Docks, 154
Easton, Alexander, 57, 84, 144, 194; and the Caledonian Canal, 96, 100; and the Bude Canal, 156; and the Birmingham and Liverpool Junction Canal, 188
Easton, George, and the Birmingham and Liverpool Junction Canal, 188
Eau Brink Cut, 161, 162
Edinburgh, Telford in, 22, 80–81
Edinburgh and Glasgow Union Canal, 165
Edleston Brook, 190, 191
Edström, Swedish engineer, 115
Effgill, 17
Egington, Francis, 42 and n.
Eil, Loch, 85, 94
Elg, Glen, 93
Elizabeth I, Queen, 129
Ellesmere Canal: preliminary plans for, 53–4; work begins on, 58;

Mersey–Dee section, 58; Llanymynech branch, 59; Wirral line, 58–9, 70; Tavern, 59; aqueducts, 64–5; tunnels, 65; difficulties and stoppages, 67 ff.; Froode Branch, 68

Ellesmere Canal Company, 49; financial difficulties, 67, 71

Ellesmere and Chester Canal, 70, 182

Elliot, John, 24

Elvanfoot Bridge, 89

English and Bristol Channel Ship Canal, 159

Erskine, Sir David, Bt., 130; and opening of the Menai Bridge, 140

Esclusham, Mt., 68

Eskdale, Telford's love of, 17, 19–20

Ettrick Pen, 17

Everleigh, John, 41

Exchequer Loan Commission, 1817, 156

Eyton, Thomas, 61

Fairness Bridge, 86

Falkirk, 89

Falkirk–Carlisle Road, 89

Fassifern, 101

Fens, drainage projects in, 161–2

Fenny Stratford, road improvements at, 126

Field, Joshua, 170, 200

Findhorn, River, 79

Fleet, Loch, 85

Fletcher, J., 158

Fochabers, 79

Forfeited Estates: Commissioners of, 83, 92; Fund, 83

Forres, 79

Forsvik, failure of coffer-dam at, 120

Fort Augustus, 78, 94, 101, 102, 104, 105; difficulties in lock-building at, 102; trouble with bottom lock, 105

Fort George, 78, 80

Fort William, 82, 85, 96

Forth and Clyde Canal, 79, 165

Fortrose, harbour improvements at, 89

Fosters Booth, road improvements at, 126

Foster, John (later Baron Oriel), 124, 125

Fradley Junction, 50

France, war with, 95. *See also* Napoleonic Wars

Frankton Junction, 70

Fraser, Alexander, 114

Fraser, Daniel, 118

Frazerburgh, harbour improvements at, 89

Freeth, John, 180, 196

French Revolution, influence of, 33–4

Frugarden Manor, 110

Fuller, A. E., 140

Fulton, Henry, 109

Fuseli illustrations, 44

Galton Bridge, 181

Garrow, Robert, 85, 89

Garry, Glen, 92–3

Garry, Loch, 93

Garry, River, 92–3

Gibb, Sir Alexander, 39, 64

Gibb, John, 88, 148 n., 162; and Dean Bridge, 150

Gilham's Coffee House, 201

Gill and Hodges, 129

Gilpin, Thomas: *Observations Relative to Picturesque Beauty*, 35

Glasgow, new water supply scheme for, 76

Glasgow–Berwick, survey for railway, 167

Glasgow, Paisley and Ardrossan Canal, 165

Glencoe, Pass of, 78

Glendinning, 17, 19

Glenfinnan, 85

Gloucester and Berkeley Ship Canal, 157–9

Gloucester, Over Bridge, 148–50

Goethe, poems of, 75

Golborne, John, 130

Goldenhill Colliery, 173

Gooch, Tom, 169 and n.

Göta, River, 108

Gotha Canal: preliminary request for, from King Gustav Adolf, 107; commercial and strategic reasons for, 108; building of, 108 ff.; proposed eastern line of, 111; work begun, 112; labour difficulties, 112, 115–16; cost, 117–18; equipment from England, 118–19; western line completed, 120; further difficulties, 120–21; opposition, 121; formal opening, 121

Gotha Canal Company, launching of, 112

Gothenburg, 108, 109, 110

Grampians, the, 78

Grand Junction Canal, 164 n., 171

Grand Trunk Canal, 50; authorizing Act, 50

Grand Union Canal, 164 and n., 175, 182

Gread Bedford Level, 161

Great Glen, the, 92 ff.

Great Haywood, 50, 185

Great North Road: surveying of, 145; Telford's recommendations for, 145

Great Ouse, River, 161

Great Soudley, 189

Greaves, John, 168

Green, James, 157

Gregory, Professor, 81

Gunthorpe, 161

Gustav Adolf, King, of Sweden, 107, 111

Guyhirn, 162

Gwynn, Captain Mark, R.N., 94

Hack Green locks, 189

Hackworth, Timothy, 172

Hadfield, Charles: *Canals of Southern England*, 157

Hamilton Bridge, 89

Hamilton, George, 178

Harbour improvements, in Highlands, 88 ff.

Hardings Wood, 179

Hardwick, Philip, 156

Hardwicke, 157

Harecastle Hill, new tunnel under, 50, 173, 181–2; inspection of old tunnel, 173–4; rebuilding process, 175–7

Harris, Telford's church on, 89

Harrison, Thomas, 178

Hatton, a mason, 26

Haw Bridge, 146, 148 n.

Haycock, J. H., 36

Hazledine, John, 66

Hazledine and Thompson, 118

Hazledine, William, 66–7, 69, 71, 72, 73, 87, 128, 134, 135, 194; his Coleham works, 135; and opening of the Menai Bridge, 140, 141; and Mythe Bridge, 148

Hebrides, the, 82

Helmsdale Bridge, 86

Hetton Colliery Railway, 171

High Offlley, 190

Highland Railway, 85; Dornoch branch line, 85

Highland Roads and Bridges, Commissioners for, 82–3

Highland Society, 78

Highlands, the. *See* Scotland

Hockcliffe road improvements at, 126

Holt Fleet Bridge, 148

Holyhead, proposed harbour works at, 124

Holyhead Road Commission, 125, 132, 133

Hope, James, 83

Hourn, Loch, 93

Howard, John, 35–6

Howth: harbour improvements at, 124, 144

Howth–Dublin Road, improvements in, 144

Huddart, Captain Joseph, 123, 125, 161

Hughes, William, 109

Hurleston, 70

Hurstmonceux Place, 28

Institution of Civil Engineers, 75, 205, 206; formation of, 200–202

Inverness, 78, 79, 82
Iona, Telford's Church on, 89
Ireland: harbour improvements in,
144; Telford's opinion of, 145
Irenant slate quarries, 70
Islay, Telford's Church on, 89
Islington tunnel, 175

Jackson, Janet. *See* Telford, Janet
Jackson, Thomas (cousin), 20, 23
Jackson, Mr. (uncle), 19
James, William, 168
Jessop, William, 54, 56–7, 61, 62, 74,
94, 153, 154; and the Cysyllte
aqueduct, 63–4; and the Chirk
aqueduct, 66; and Menai Straits
suggestion, 130; and the West India
Docks, 154
Johnstone, George, 44
Johnstone, Sir James, 23, 27
Johnstone mausoleum, the, 18
Johnstone, William. *See* Pulteney
Jolliffe and Banks, 162
Jones, Inigo, 41
Jones, James, 200
Jones, William, 75
Jordan, Mrs., 46

Kendal Coffee House, 201
Kendal tunnel, 175
Ketley: road improvements at, 126
Ketley Canal, 52, 60
King's Lynn, 161
Kinloch Ailort, 85
Kirkwall, harbour improvements at, 89
Knighton, 189, 190
Knipersley Reservoir, 177, 178
Knoidart, 82
Kyleakin ferry, 85
Kynaston-Powell, John, 53, 54

Labelye, 199
Lagerheim, Swedish engineer, 115,
119, 122
Laggan summit cutting, 102–3; slips
in, 105

Lanarkshire, new road through, 89
Lancaster Canal, 175
Landhöjden, 111
Langholm, 20–22, 24, 75
Langholm Patriots, 34
Lawton Locks, 176
Lee, Thomas Eyre, 182
Leicestershire and Northamptonshire
Union Canal, 164 n.
Lewis, Telford's church on, 89
Leys Ironworks, 99
Linnhe, Loch, 100, 101
Little, Andrew, 19, 203; blinded by
lightning, 24; letters from Telford,
26, 28, 29–31, 32, 34, 36, 40, 55, 58,
60–61, 74, 80, 81, 198; receives
money from Telford, 43
Little, James, 199; on life in Abingdon
Street, 199
Little, William, 19, 24, 199, 203;
letters to, 144
Liverpool: Telford on, 76; new water
supply scheme for, 76
Liverpool and Manchester Railway,
168–70, 187, 204
Liverpool Ship Canal, 159
Llangollen, Vale of, 54, 59, 62, 70, 71,
124
Llanrwst: Denbigh Bank at, 134
Llantisilio, 70
Lloyd, Rev. John Robert, 53
Lochaber, 80
Lochalsh, 82
Lock, James, 194
Locke, Joseph, 168
Lockgate Bridge, 59
Lochy, Loch, 92, 94, 101, 102
Lock-building, 98–9
Lomond, Loch, 78
London: Highgate Archway cutting,
126; Docks and Dock Companies,
154–5
London Bridge: rebuilding of, 76;
Telford's design for, 153; Rennie's
design, 154, 159
London and Birmingham Junction
Canal scheme, 182

London–Chepstow mail coaches, 146–7
London–Holyhead: road survey, 125; building of new road, 126 ff.; cost of, 126
London–Shrewsbury Road: turnpikes on, 124; Telford's proposals for, 126
Longdon, 61
Longport, 176
Lovat Bridge, 86
Lowlands. *See* Scotland
Loy, River, 101
Lyon, David, of Aberdeen, 114

Macadam, John London, 126, 127
Macclesfield, 179; Canal, 178–9
Macdonald, Alexander, of Inverness, 115
Macfarlane, George, 85
McInnes, Daniel, 85
McIntrop, Hugh, 148
McKenzie, Alexander, of Strathpeffer, 114
Madeley Church, 40, 42, 58
Maida Hill tunnel, 175
Malcolm, Admiral Sir Pulteney, 200
Mallaig, 85
Manchester Ship Canal, 159
Market Drayton, 189, 190
Marple, 69, 179
Marquand, J., Admiralty Surveyor, 28
Maudslay, William, 200
Megget Water, 17, 18
Meikle Ferry, 88
Mem, 111, 121
Menai Bridge: opposition to in Caernarvon, 133; work begins on, 133; raising of first chain, 137–8; opening of, 139–40; and storms, 151–2
Menai Straits: problem of crossing, 129; Pig Island, 130, 131, 133; Swilly Rocks, 130, 131
Meriden, road improvements at, 126
Mersey, River, 50, 53, 131
Mersey and Irwell Navigations, 171
Middlewich, 53, 185

Milford Haven–Waterford steam packet service, 145
Mills, James, 168–9, 182
Milton, John, 22
Mitchell, John, 84–5, 86; narrow escape, 88
Mitchell, Joseph, 85; and church construction in Highlands, 89; and South Wales road survey, 147; on life in Abingdon Street, 199–200; on Telford's deafness, 202
Monckton, Edward, 194
Montfaucon: *Antiquities*, 35
Montford Bridge, 47
Montgomeryshire Canal, 58, 70
More, Sir Thomas, 82
Morpeth Bridge, 145
Morriston, Glen, 94
Motala, 111, 112, 118, 119; Motala Verkstad, 118
Mound, Station, 85
Mouse Water, 90
Mull, Telford's Church on, 89
Muirtown, 105; lock-building at, 99
Murray, Robert, 85
Myddleton, Richard, 64
Mylne, Robert, 157, 159
Mythe Bridge, 148, 160

Nant Ffrancon Pass, 123, 125, 128, 129
Nantwich, 53, 188; settling of embankment, 191
Napoleonic Wars, 105, 106, 165
Naseby Wolds, 165 n.
Navigation Commission, 161
Nelson engine, the, 175, 176
Nene, River, 161, 162
Neptune's Staircase, 101, 123
Ness, Loch, 92, 94, 99–100
Ness, Loch–Lochalsh, Kyle of, Road, 85
Netherpool–Chester canal section, 58
New Hall Colliery, 68
New Passage Ferry, 146
Newcastle and Carlisle Railway, 168
Newport, 188
Newtown, 58
Norbury, 188, 190

Norham Ford Bridge, 132
North British Railway, 85
Northleach–Carmarthen mail coach road, 145
North Level, draining of, 162
North Ronaldshay, Telford's church on, 89
North Uist, Telford's church on, 89

Observations on Railways with Locomotive high-pressure Steam Engines, 166
Ogwen, Lake, 125
Ogwen, River, 129
Oich, Loch, 92, 94, 102
Oich, Loch-Hourn, Loch, Road, 85
Old Stratford, road improvements at, 126
Owen, W. Mostyn, 53, 54
Oxford: Telford in, 41–2; Trinity Chapel, 42; Christchurch, 41–2; All Saints Church, 42
Oxford Canal, 50, 182
Oxford Canal Company, 50
Oxonian, London stage coach, 140, 151
Oykell, River, 88

Padarn slate quarries, 133
Paine: *Rights of Man*, 33–4
Palmer, Henry R., 171, 200
Palmer, John, 41
Palmer, W. R., 182
Parnell, Sir Henry (later Lord Congleton), 125, 132, 133, 200; and opening of Menai Bridge, 140
Pasley, General, 170, 200
Pasley, James, 19
Pasley, John, 25
Pasley, Miss, of Langholm, 21–2, 25
Pathhead, Bridge, 150
Peak Forest Canal, 179
Peak Forest Tramway, 69, 167
Penmaenmawr, 124
Penmon, limestone from, 133
Penrhyn Castle, 28
Pentland Firth, 94, 105
Perronet, French architect, 149

Peterhead, 80; harbour improvements, 89
Pickering, Exuperius, 68, 69
Piff's Elm, suggested new road from, 146
Pilot, Bangor stage coach, 140
Pitt, William, 79
Plas Kynaston, foundries at, 66–7, 101
Platen, Count Admiral Baltzar Bogislaus von, 132, 134, 142; first meets Telford, 110; correspondence with Telford, 112–13, 114, 159, 174, 177, 180, 187; obtains English tools, 112–13; trouble with Scottish workers, 116–17; appeal for additional canal grant, 117–18; and loss of Bagge, 115; comes to England, 119; dies of cancer, 121; Telford requests material from, for library of Institution of Civil Engineers, 201–2
Playfair, Professor, 81, 153
Plymley: *General View of the Agriculture of Shropshire*, 75, 167 n.
Poetry, Telford's delight in, 22, 29–30
Pollok, John, 89
Ponkey Colliery, 68
Pont Cysyllte, 123; building of aqueduct at, 59–60, 61; foundation stone laid, 64; ceremonial opening, 72
Portsmouth Dockyard, Telford's work at, 28
Potarch Bridge, 86
Pot Shrigley, 179
Potter, James, 174, 176–7
Preston Brook, 50, 185
Price, Henry Habberley, 146
Priors Lee, road improvements at, 126
Prison reform, 35–6
Pritchard, Daniel, 175
Pritchard and Hoof, 174–5, 176 179
Pritchard, Thomas Farnolls, 48
Provis, John, 135; and opening of Menai Bridge, 140; and opening of Conway Bridge, 141

Provis, William, 57, 90, 125, 179, 190, 191, 192, 193, 201; and work on Menai Bridge, 133, 137; opening of Bridge, 140–41; account of, to Telford, 141–2; and Birmingham and Liverpool Junction Canal, 188

Pulteney, William, 27, 58, 77; patronage of Telford, 32; character of, 33

Quarff, Telford's church on, 89
Quoich, Loch, 93

Raatachan, Mt., 93
Railways, influence of, on canal engineering, 69; effect on canal companies, 165 ff.
Rastrick, J. U., 168
Read, William, 140
Regent's Canal, 164, 165 n., 175
Rennie, John, 28 n., 69, 81, 88, 92, 107, 123–4; and Menai Straits suggestion, 130, 131; and London Docks, 154; and East India Docks, 154; and London Bridge, 154; death, 156, 159; wife's death, 160; relations with Telford, 159–60; as bridge-builder, 160–61; as canal engineer, 161; Fen drainage work, 161; report on Harecastle Hill tunnel, 173–4
Rennie, Sir John, 161, 162, 168, 206
Reynolds, Sir Joshua, 42
Reynolds, William, 52–3, 55, 60, 61, 67, 153
Rhodes, Thomas, 102; and work on Menai Bridge, 136, 137; opening of Bridge, 140; account of storms striking, 151; and St. Katherine's Dock, 155
Rickman, John, 83, 84, 89, 105, 162, 170, 181; and the Caledonian Canal, 96
Road making: in the Highlands, 85 ff.; Macadam's, 126; Telford's, 126–7
Robison, Professor, 81 and n., 153
Rochester tunnel, 175

Rödesund, 111, 120
Rotton Park reservoir, 181
Rowland, Edward, 68, 69
Roxen, Lake, 111, 121
Royal George locomotive, 172
Royal London and Holyhead Mail, 140
Ruabon, 53, 54, 59, 68; Tramway, 165–6
Rubens, Peter Paul, 42
Runcorn, Bridge, designs for, 131–2

St. David, the, 138
St. Katharine's Dock Company, 154
Salopian Coffee House, 198–9
Sarsfield, D., 138
Sätra, 111
Sätraån, River, 111
Saumarez, Admiral Sir James, 109
Schweder, Elias, 109
Scotland:
  The Highlands: eighteenth-century history of, 77 ff.; effects of destroying clan system, 77–8; military roads, 78; Telford on causes of emigration from 82; making of new roads, 85–6; bridge-building, 86–8; harbour improvements, 88–9; church and manse building, 89
  The Lowlands: Industrial Revolution in, 79; communications in, 79; bridge and road construction in, 89–90
Scott, Sir Walter, 30–31, 60
Severn, River, 50, 53; as eighteenth-century Shropshire trade route, 52; improvement of navigation, 76; bridges, 47–9, 58, 148–50
Sharpness Point, 158, 159
Shebdon, village of, 190
Sheep Island, 101
Shelmore Great Bank, 190; trouble at, 192–7
Sheridan: Biography of Jonathan Swift, 46
Shiel, Glen, 85, 94

Shiel, Loch, 85, 94

Shin, the, 17

Shrewsbury: Castle, Telford goes to, 31; his work in, 32 ff.; construction of new infirmary and county gaol, 35; affair of St. Chad's Church, 37–8; St. Mary's Church, Telford's pulpit in, 38; foundries, 66

Shrewsbury Canal Company, 60

Shrewsbury–Holyhead Road: turnpikes on, 124; Telford's proposals for, 126

Shropshire bridges, 58

Shropshire Canal, 52–3

Shropshire Union Railways and Canal Company, 188

Simpson, James, 114, 118; bad behaviour of, in Sweden, 116

Simpson, John, 47, 48, 65, 70, 73, 87, 134, 158; work on the Caledonian Canal, 96; death, 106, 188

Sinclair, John, 125

Sinclair, Sir John, 33

Sjötorp, 111

Skye, Telford's church on, 89

Smeaton, John, 34, 54, 79, 88, 206

Smethwick, 180

Smiles, Samuel: *Life of Telford*, 20, 22, 38, 131, 153, 198; on Rennie, 161

Smirke, Sir Robert, 149 and n.

Smith, Adam: *Wealth of Nations*, 44, 75

Smith, Asheton, 133

Smith, James, 85, 114

Smith, Jenny, 24

Society of Civil Engineers, 34, 200

Söderköping, 121

Solihull, 182

Somerset House, Telford's work on, 25–6

South Mimms, road improvements at, 126

South Wales, road survey, 147

Southey, Robert, 88; on Telford, 90–91; on Banavie locks, 101; on Laggan summit cutting, 103–4; on Telford's roads, 126–7; and 'the colossus of roads', 142; and

'Pontifex Maximus', 142; in Abingdon Street, 200; legacy from Telford, 205

Spence, Thomas, 85

Spey, River, 79; bridge over, 83

Sproat, Robert, 125

Stafford, Marquis of, 194

Staffordshire and Worcestershire Canal, 50, 180, 194 n.

Stanley Sands, embankment across, 129

Stanton, Thomas, 72, 167

Steamers, paddle, on Caledonian Canal, 105; for Gotha Canal, 118–19

Stephenson, George, 166, 168, 169

Stephenson, Robert, 206

Stevenson, John, 100

Steuart, George, 38, and n.

Stewart, Professor Dugald, 80; *Philosophy of the Human Mind*, 75

Stockton and Darlington Railway, 165–6, 171–2

Stone, quarrying of for lock-building, 99, 101

Stourport, 50

Straphen and Hall, 133

Stratford Canal, 182

Stratford and Moreton Railway, 168

Strath Bran, 85

Strath Dearn, 79

Stromness, 95

Stuart and Revet: *Antiquities of Athens*, 35

Stuttle, William, 87, 128

Sudborough: Telford's alterations to vicarage at, 27

Sutherland, 82, 85

Sutton, 179

Swansborough, William, of Wisbech, 162

Sweden, 1809 Revolution in, 111–12

Tanat, River, 124

Tay, River, 79; bridge over, 83, 87

Taylor, Joseph, 115, 154

Telford, Janet (mother), 18, 19; last year, 75

Telford, John (father), 17; death, 18

Telford, John (kinsman): work on Caledonian Canal, 95–6, 100

Telford, Thomas: birth, 18; childhood, 19; schooldays, 19; love of Eskdale, 19–20; mason's apprentice, 20; introduced to English literature, 22; stay in Edinburgh, 22; to London, 22; becomes first-class mason, 25; work for William Pulteney, 27; work on Portsmouth Dockyard, 28; self-education, 29; delight in poetry, 29–30; as Surveyor of Public Works for County of Shropshire, 32; revolutionary sympathies, 33–4; excavations of Uriconium, 36–7; his autobiography, 39, 47, 203; *Atlas*, 39; 159; to Bath, 40–41; to London, 41; to Oxford, 41–2; to Birmingham, 42–3; sends money to his mother and Little, 43; poetry writing, 43–4, 74–5; as a reader, 46, 75; supporter of theatre, 46; early bridge-building, 47; with Ellesmere Canal Company, 49, 55–6; nature of his appointment, 56–7, abandons architectural work, 58; Engineer to the Shrewsbury Canal, 60–61; his plans for Pont Cysyllte passed, 62; work begun, 64; work on Chirk aqueduct, 65–6; restoring of Chester Canal, 70; *Treatise on Mills*, 75; writing on canals, 75; love of countryside and agriculture, 76; survey of Highland communications, 77–8, 80; further visit, 81–2; communications scheme, 82; made Fellow of Royal Society of Edinburgh, 82; Scottish road construction work, 84, 85–6, 89; bridge-building, 86–8; harbour improvements, 88–9; church-building, 89; survey for Caledonian Canal, 92 ff.; estimate for, 94; building of, 94 ff.; peak of his fame, 107; invited to Sweden for Gotha Canal project, 107, 109; sails, 109–10; meets von Platen, 110;

surveys canal line, 110; report on, 111; created Knight of Royal Order of Vasa, 113–14; invites von Platen to Britain, 119; his work on the Gotha, 122; surveys London–Holyhead route, 124; Parliamentary report on his findings, 125; suggested improvements, 125 ff.; problem of Menai Straits crossing, 130–31; Runcorn Bridge project, 131–2; fresh design for Menai Bridge 132; work begins, 133; method of paying men, 134; worries over, 136–7; his opinion of his work, 142; modern view of, 142–3; popularity of, 143; his interest in fellow-men, 143; travels, 143–5; road surveying, 1820–1830, 145; Great North Road survey, 145; South Wales road survey, 145–7; further bridge-building, 148 ff.; and proposed Clifton Bridge, 152; work on London Docks, 154 ff.; worried by hasty work, 155; engineering adviser to Exchequer Loan Commission, 156 ff.; Gloucester and Berkeley Ship Canal, 157 ff.; Fen drainage work, 161–2; beginnings of failing health, 162–3; views on railways, 167; inspection of Harecastle Tunnel, 174; Macclesfield Canal survey, 178–9; Birmingham Canal survey, 179–80; Birmingham and Liverpool Junction Canal, 184 ff.; further illness, 194; death, 196; life at the Salopian Coffee House, 198–9; move to Abingdon Street, 199; first President of Institution of Civil Engineers, 201; Fellow of Royal Society, 202; deafness in later years, 202–3; statue in Westminster Abbey, 203; bequests, 205

Telford, William (kinsman), 44

Tern, River, 61, 62, 190

Tewkesbury, 148, 160

Thames and Medway Canal, 175

Thomson, Andrew, 20–21
Thomson, James, of Glasgow, 118
Thomson, James, poet, 75
Thomson, Tibby, 21
Thorney, 162
Thunberg, Daniel av, 109
Thurso, 85
Tipton locks, 181
Tobermory, 79
Tomkinson, Mr., of Dorfold Hall, 189
Tongue, 85
Tongueland Bridge, 89–90, 138
Torridon, Lake, 85
Tower Bridge, 139
Tref-y-Nant brook, 71
Trench, 60
Trent, River, 50
Trent and Mersey Canal, 50
Trent and Mersey Canal Company, 50, 53, 164; and Harecastle Tunnel, 173–5; opposition to Birmingham and Liverpool Junction Canal Company, 185, 187
Trevithick, Richard, 66
Trollhätte Canal, 108, 109, 116, 122
Trubshaw, James, 178
Turner, William, 61–2
Turnpike Trusts, 124, 126, 170
Tweed, River: Union Bridge over, 132
Tyndrum, 78
Tyrley locks, 189

Ullapool, 79
Ulva, Telford's Church on, 89
Unden, Lake, 111
Upton Forge, 135
Upton, John, 158
Uriconium, excavation of, 36–7
Urquhart, William, of Inverness, 114

Valle Crucis, 70
Valley village, 129
Vänern, Lake, 108, 110, 111, 119
Vansittart, Nicholas, 80, 156
Varley, James, 62, 65
Vättern, Lake, 108, 110, 111, 120

Vaughan, William, 178
Victory, the, 109
Vignoles, Charles, 168
Viken, Lake, 111, 119, 120
Vron Cysyllte, 69, 71, 72

Walker, Ralph, 154
Walpole, Horace, 45
Wansbeck, River, 145
Wappenshall, 189
Warehousing Act, 1823, 154
Wash, the, 161; Crab's Hole Sluice, 161
Watt, James, 28, 81, 92, 153, 160
Weaver Navigation Trustees, 165
Weaver, River, 190
Wedgwood, Josiah, 178
Welch, Henry, 147
Welshpool, 58
Westerhall, Telford's work on, 27
Westerker Rig, 17
Westerkirk, 18
West India Docks, 154
Westminster Bridge, 199
Weston Canal, 165
Weston Lullingfields, 70; Ellesmere Canal wharf at, 135
Weston, Samuel, 54, 58
Wheaton Aston, 189
Whitchurch, 70
Whitehouses tunnel, 65, 177
Whitworth, Robert, 79
Wick, 85; Bridge, 86
Wilden Ferry, 50
Wilkinson, John, 56, 67, 81 n., 153
Williams, John, owner of Menai Ferry rights, 129
Williams, John, a carpenter, 138
Williams, Miss, of Plas Isa, 130
Williams, Owen, of Craig-y-Don, 133
Williams, William, labourer, 138
Wilson, John, 65, 73, 114, 136, 189, 190–91; and work on Caledonian Canal, 96, 102; trouble in Sweden, 116–17; and Menai Bridge, 134, 137, 151; and Birmingham and Liverpool Junction Canal, 188; death, 192
Wing, Tycho, 162

Wolverhampton–Mersey Canal
    scheme, 182
Woodhouse, John, 158
Woodseaves, 191
World War I: North Sea mine
    barrage, 106
Wrekin, the, 52
Wren, Sir Christopher, 41, 42
Wrexham, 53, 54
Wrottesley, Sir James, 194

Wroxeter, excavations at, 36
Wyatt, James, 28, 35, 39, 41,
    42
Wyatt, Samuel, 28
Wynn, Sir Watkin Williams,
    71
Wyrley and Essington Canal
    Company, 194 n.

Yeats, William Butler, 45